TIME LAPSE APPROACH TO MONITORING OIL, GAS, AND CO_2 STORAGE BY SEISMIC METHODS

TIME LAPSE APPROACH TO MONITORING OIL, GAS, AND CO_2 STORAGE BY SEISMIC METHODS

JUNZO KASAHARA

Tokyo University of Marine Science and Technology, Shizuoka University

YOKO HASADA

Daiwa Exploration and Consulting Co., Ltd.

AMSTERDAM • BOSTON • HEIDELBERG • LONDON
NEW YORK • OXFORD • PARIS • SAN DIEGO
SAN FRANCISCO • SINGAPORE • SYDNEY • TOKYO
Gulf Professional Publishing is an imprint of Elsevier

Gulf Professional Publishing is an imprint of Elsevier
50 Hampshire Street, 5th Floor, Cambridge, MA 02139, United States
The Boulevard, Langford Lane, Kidlington, Oxford, OX5 1GB, United Kingdom

Copyright © 2017 Elsevier Inc. All rights reserved.

No part of this publication may be reproduced or transmitted in any form or by any means, electronic or mechanical, including photocopying, recording, or any information storage and retrieval system, without permission in writing from the publisher. Details on how to seek permission, further information about the Publisher's permissions policies and our arrangements with organizations such as the Copyright Clearance Center and the Copyright Licensing Agency, can be found at our website: www.elsevier.com/permissions.

This book and the individual contributions contained in it are protected under copyright by the Publisher (other than as may be noted herein).

Notices
Knowledge and best practice in this field are constantly changing. As new research and experience broaden our understanding, changes in research methods, professional practices, or medical treatment may become necessary.

Practitioners and researchers must always rely on their own experience and knowledge in evaluating and using any information, methods, compounds, or experiments described herein. In using such information or methods they should be mindful of their own safety and the safety of others, including parties for whom they have a professional responsibility.

To the fullest extent of the law, neither the Publisher nor the authors, contributors, or editors, assume any liability for any injury and/or damage to persons or property as a matter of products liability, negligence or otherwise, or from any use or operation of any methods, products, instructions, or ideas contained in the material herein.

Library of Congress Cataloging-in-Publication Data
A catalog record for this book is available from the Library of Congress

British Library Cataloguing-in-Publication Data
A catalogue record for this book is available from the British Library

ISBN: 978-0-12-803588-7

For information on all Gulf Professional Publishing visit our website at https://www.elsevier.com/

Working together to grow libraries in developing countries

www.elsevier.com • www.bookaid.org

Publisher: Candice Janco
Acquisition Editor: Amy Shapiro
Editorial Project Manager: Tasha Frank
Production Project Manager: Maria Bernard
Designer: Mark Rogers

Typeset by TNQ Books and Journals

CONTENTS

PREFACE

In recent years oil and gas industries entered a new stage called unconventional natural resources. The unconventional natural resources are heavy oil reservoirs such as oil sands and shale oil, shale gas, deep-sea oil and gas fields, coal-bed methane, and oil and gas in sub-basalts. The methane hydrate will be an unconventional natural resource in the future. To produce heavy oil efficiently, enhanced oil recovery (EOR) technology has been used. In EOR, vapor or supercritical CO_2 is injected to the heavy oil layer to increase the mobility of heavy oil by softening and decreasing viscosity. During injections, it is better to monitor the physical state of the reservoir by seismic methods in addition to reservoir engineering technologies using history matching. The monitoring of physical state changes in EOR is called "time-lapse" technology. Carbon capture and storage (CCS) uses the same technology as CO_2-EOR.

The time-lapse monitoring during EOR has been done by various methods: repeated 3D seismic surveys called 4D seismic survey, well loggings, distributed acoustic sensor, well-to-well seismic or resistivity tomographies, vertical seismic profile, interferometry synthetic aperture radar, distributed temperature sensor at borehole and/or surface, passive seismic method to monitor earthquakes, seismic interferometry, and Accurately Controlled and Routinely Operated Signal System (ACROSS) seismic method.

The permanent reservoir monitoring is one of the time-lapse technologies. This technique has been extensively used in the North Sea region, off Azerbaijan, and off Brazil by using ocean bottom cable, or ocean bottom seismograph laid down at the ocean floor. In these fields the monitoring of existing reservoirs has been carried out to enhance the efficiency of oil recovery. The shootings from the ocean surface have been repeated to learn the physical states of oil fields. Time lapse can tell the temporal change of oil reservoirs.

During shale gas production, passive seismology technique is used to monitor the fracking quakes. The 4D seismic and other methods are not commonly used because the average life of each well in shale gas production is short, and these measurements raise the production costs. If the monitoring costs becomes cheaper, the time-lapse technology can be used in shale gas production.

In this book, a new ACROSS time-lapse technology is introduced. This method is relatively new for the geophysical exploration tools. ACROSS is based on signal-processing strategy developed by Dr. Kumazawa and his colleagues since 1994 just before the 1995 Great Hanshin earthquake (Mw = 6.9) in Kobe, Japan, and it has been applied to seismic measurements and electromagnetic measurements. He intended to use this technology for the continuous monitoring of earthquake nucleation processes along the plate subduction boundaries. The authors have spent their efforts applying this technology to the time-lapse approach for oil and gas exploration and CCS.

The ACROSS seismic source is comprised of motor and eccentric weight. By the rotation of an eccentric weight by a motor, the centrifugal force is generated. The instantaneous position of weight mass is controlled in reference to the very accurate GPS time standard. The ACROSS seismic source uses a similar frequency sweep method as one used in conventional vibrators, except for the concept of constant repetition. For example, the sweep from f_1 to f_2 during the repetition time frame of T_m generates a set of line spectrum from f_1 to f_2 with the frequency spacing of $1/T_m$. Because the line spectra of source signature are precisely obtained, the enhancement of signal-to-noise ratio (S/N) could be achieved by stacking of plural sets of sweeps in time domain or frequency domain. If the source signature does not change during several days, you can stack the data during several days. One of the ACROSS seismic sources can generate 3.9×10^5 N at 50 Hz comparable to the large land vibrator, but stacking the long data can give much higher S/N over 10–50 Hz than ordinary seismic vibrator source.

The repeatability of time lapse is the most important factor. By mounting the ACROSS seismic source on the heavy concrete base, we can obtain high repeatability even if sources and geophones are at the surface. By use of buried geophones at a few tens of meters, much better repeatability can be achieved.

Other characteristics are simultaneous generation of vertical and horizontal forces in the case of the horizontal rotational axis. The addition of received signals from clockwise and counterclockwise rotations gives the vertical force response, and the subtraction of two received records from both rotations gives the horizontal force response. The P- and S-waves are dominant in the former and latter processed records, respectively. The simultaneous excitations of vertical and horizontal vibrations cannot be achieved by conventional seismic sources.

Another uniqueness of the ACROSS seismic source is the separation of background noises and source signals. If the fracking quakes in shale gas production are included in background noises, you can separate fracking quakes from active seismic signals. By this technique, the active and passive simultaneous time-lapse methods can be obtained. If the noises come from traffic noise, human activities, and natural noises generated by weather changes, the cross-correlation of separated noise components between two locations gives seismic interferometry, which gives Green's functions between two locations, and the monitoring of seismic interferometry can be used for the time-lapse study.

Because the ACROSS seismic source is placed to solid base minimizing the temporal change of source signature to obtain good repeatability, it is difficult to get dense source spacing. In order to make sure to obtain reasonably good imaging by a few fixed-source locations, simulations for this circumstance and field tests have been carried out. The several simulations using reverse-time technique gave good images of temporary changing zones even if only one seismic source and array of geophones was used. The field test was carried out by air injection to subsurface, and the migration of air in the subsurface was imaged time to time. Currently the ACROSS technology has been tested in Aquistore, Saskatchewan, Canada, in CO_2-EOR experiment.

This book summarizes the time-lapse studies in the world and describes the technology of ACROSS time lapse with the principle of ACROSS methodology and unique processing methods.

We hope that this book helps you understand the new ACROSS technology and that you will apply the ACROSS time-lapse technology to oil and gas and CCS explorations.

ACKNOWLEDGMENTS

The Accurately Controlled and Routinely Operated Signal System (ACROSS) technologies have been developed in the Tono Geoscience Center of the Japan Atomic Energy Agency by Drs. M. Kumazawa, T. Kunitomo, T. Nakajima, K. Tsuruga, H. Nagao, and Y. Yokoyama and in Nagoya University by Drs. K. Yamaoka, T. Watanabe, and R. Ikuta. Dr. N. Fujii helped them though aggressive discussion. The authors express our great thanks to them for their continuous effort to develop the ACROSS methodology. Dr. T. Kunitomo gave us a great deal of advice for manufacturing and operating ACROSS seismic source. He made the hardware and software designs of ACROSS seismic source.

We also express our great thanks to JCCP (Japan Cooperation Center Petroleum) and officers of JCCP for their financial support and their aggressive support of our studies and field surveys in Saudi Arabia. We express our great thanks to many colleges in Saudi Arabia and Japan, including Dr. K. Al-Damegh; Messrs. G. Al-Aenezi, K. AlYousef, O. Lafouza, F. Almalki, A. Alhumaizi, I. Alrougy, M. S. Alajmi, I. Ajurayed (KACST); Drs. R. Kubota, K. Murase, A. Kamimura, and Messrs. O. Fujimoto, H. Ohmura, E. Nishiyama, Y. Kanai, O. Tazawa, G. Kato, T. Hasegawa, H. Fukatsu, and Ms. Y. Mori (Kawasaki Geological Engineering Co. Ltd.), and Mr. S. Ito, Dr. A. Guidi, Messrs. M. Takano, and T. Fujiwara (NTT data CCS Co. Ltd.) for their great efforts on fieldwork and data processing. Without their great efforts, we could not have obtained the present results.

Mr. K. Ino of Sanko Keisoku Service Co. Ltd. and engineers in Tomei-Koki Co. Ltd. manufactured ACROSS seismic source and gave useful advice about its operation.

We express our great thanks to Ms. Tasha Frank, Marisa LaFleur, Amy M. Shapiro, Maria Bernard, and Mr. J. Fedor for encouraging us to publish this work and assisting our writing.

Finally, our thanks go to Ms. K. Sekiguchi who helped with the final stage of writing.

CHAPTER 1

What is Time Lapse?

Contents

1.1 INTRODUCTION

Although the main objectives of this book are to introduce the general outline of the time lapse, this book also has unique objectives. This book describes a **unique signal processing method** based on the principle of the Accurately Controlled and Routinely Operated Signal System (ACROSS) methodology. By this methodology, the signal generated by a control source can be separated from the background noise. For example, during the operation of control source(s), the background noise such as traffic noise, human activity noise, acoustic emissions by fracking, etc. can be separated from signals from control sources(s) at the same circumstances. By applying the processing of separation of signal from noise, the signal-to-noise ratio (S/N) can be quantitatively determined. By increasing data length in stacking, you can see the S/N increases; the signal level stays constant and the noise level decreases (Fig. 1.1) (see Appendix C.3). By the method of separation of signal from noise the vibration of ACROSS seismic source does not disturb other signals, even if a geophone is a neighbor of a vibrator. Furthermore, plural ACROSS vibrators can be operated at the same time without mixing each other's signals. If you can employ the frequency sweep used in ACROSS and processing similar to ACROSS methodology, you can run plural vibrators by different sweep patterns (see Appendix C.2).

Time Lapse Approach to Monitoring Oil, Gas, and CO_2 Storage by Seismic Methods
ISBN 978-0-12-803588-7
http://dx.doi.org/10.1016/B978-0-12-803588-7.00001-7

Copyright © 2017
Elsevier Inc.
All rights reserved.

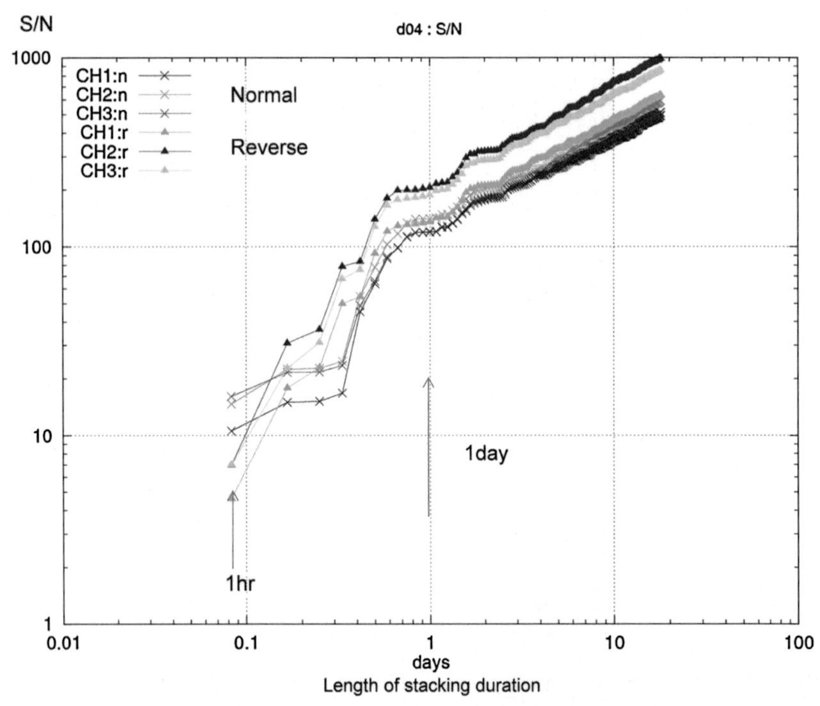

Figure 1.1 Improvement of S/N by increasing data length. Normal and reverse: clockwise and counterclockwise rotations of ACROSS, respectively. In this case, by one-day-long stacking the S/N is more than 100 on this station.

As described in this book, a repetitive sweep or chirp signal has a set of line spectra. For this reason, the enhancement of signal-to-noise ratio can be achieved by stacking on known line spectra. This is the **stacking in spectral domain** compared to **stacking in time domain**. In conventional seismic exploration method, **the stacking in space** is commonly used. Although all three can enhance S/N, the most important point in the ACROSS processing is that the weighted stacking in spectral domain and use of long data length enhance the S/N. The details of ACROSS signal processing are given in Appendix C.3.

One more thing should be noted. In this book, the terminology "transfer function" is frequently used. This is the response of any geophone to a unit force applied at the source. We also frequently use the terminology "source signature." While a transfer function means output/input characteristics of a general linear system, a transfer function in ACROSS technology is obtained by R(ω)/S(ω), where R(ω) is the frequency spectrum of received seismic waveforms and S(ω) is the applied source signature in frequency domain.

1.2 OVERVIEW OF TIME-LAPSE STUDIES

During a few decades geophysical explorations made great progress in work on the technology of 3D imaging of subsurface. One of the key technologies in 3D geophysical explorations could be the seismic reflection method. The key technology of seismic reflection methods could be deconvolution, which is the method to make reflection wavelet series to impulse series. When we can handle the data of band-limited in frequency (the windowed data in frequency domain), we always encounter the effects of frequency window problem. When we transform the data of band-limited in frequency to time domain, the data could be suffered by the band-limited frequency window. The approaches by Burg (1967, 1968), Akaike (1969a,b), and Treitel and Robinson (1981) were to minimize the window effects. An impulse in time domain is white spectra in frequency domain. When we use a seismic vibrator on land with frequency contents of 5–300 Hz, autocorrelations of source waveforms for two kinds of sweep are zero-phase wavelet with side-robe shown in Fig. 3.2. As seen in this figure, one of the deconvolution objectives is the process to eliminate the side-robe of impulse, that is, the process to eliminate the band-limited window effects. Details are described in Appendix A.3. The other objectives of deconvolution are suppressing the multiples, but this is out of scope of this book.

The second stage of key technology is **migration**. This technology focuses on diffracted waves. The efforts to use **waveform inversion** could be the most recent technology. The migration and waveform inversion are similar to backpropagation. The time-lapse imaging is enabled by backpropagation of residual waveforms (Chapter 4).

According to the advance of technologies, so-called **unconventional** oil and gas exploration have attracted oil and gas industries. Unconventional oil and gas explorations have been worked for **deep-sea reservoirs**, **heavy oil reservoirs**, **shale gas fields**, **shale oil fields**, **sub-basalt fields**, and **permanent reservoir monitoring application**. To explore the unconventional resources, the time-lapse method is needed.

1.3 OBJECTIVES OF TIME-LAPSE STUDIES

In the exploration of unconventional resources, enhanced oil recovery (**EOR**) method is the most important (e.g., Schenewerk, 2012), and time-lapse approach is essential in EOR technology. EOR is used to produce oil from heavy oil reservoirs in so many oil fields in the world (Koottungal, 2014). There are several different approaches for EOR: thermal method (steam, combustion, hot water), chemical approach (micellar-polymer, polymer, caustic/alkaline, surfactant), gas injection (hydrocarbon miscible or immiscible, CO_2 immiscible or miscible, nitrogen, flue gas) and microbial method. Among these approaches, steam injection and CO_2 miscible have been the most popular (Koottungal, 2014).

Heavy oil (tight oil) such as pitch, tar, or bitumen with high viscosity is hard to migrate and get retrieving. When the temperature of heavy oil becomes high, the heavy oil is easy to move. The injection of **steam** or **supercritical CO_2 (SCC) into heavy oil layers** is used for the production of oil from heavy oil reservoirs. SCC generates miscible state of oil and CO_2 to increase the mobility of oil. SCC injection to heavy oil seems also to enhance the production rate of oil. After the injection of steam or SCC into a layer with high contents of heavy oil, the layer becomes soft and/or liquid state. The change of physical properties by injection can be observed by changes in V_P/V_S ratio and reflectivity of heavy oil reservoir.

The time-lapse approach is also considered as the essential technology in **permanent reservoir monitoring**. Extensive surveys have been carried out in the North Sea by BP, offshore of Azerbaijan, offshore of Brazil by Petrobras, and the Gulf of Mexico by BP. Three-dimensional seismic profile suggests the physical state of oil/gas reservoir. For the permanent seismic monitoring, ocean bottom cable (OBC) at seabed is used. For seismic sources, air gun shootings similar to 3D seismic method have been repeated for the permanent reservoir monitoring. Reservoir engineers decide where the effective reservoirs for production are.

In the development of **shale gas** industry the **fracking** of shale layer could help to make free gas enclosed in the shale layer. Associated with the fracking, people can expect the change of physical properties caused by increase of fractures and high degree of anisotropy due to fracturing of rocks. The time-lapse approach in shale gas exploration can be useful, but it is not seriously considered in this application because the size of shale gas mining does not seem so large and duration of production is short.

In carbon capture and storage (**CCS**) supercritical CO_2 is injected to aquifers or porous media with seal layers in deep subsurface. CO_2 in CCS needs to be stored in subsurface, against escaping from the storage zone to other places such as in air. To make sure that the injected CO_2 does not escape from the storage zone, the time-lapse approach is considered to be mandatory. In many cases, one of the EOR technologies uses miscible or immiscible CO_2. In the case of Weyburn-Midale CO_2-EOR 20 Mt CO_2 was sequestrated into the subsurface.

1.4 BRIEF REVIEW OF PREVIOUS APPROACHES FOR THE TIME-LAPSE STUDIES

The time-lapse method has been used in the field of CCS and CO_2-EOR, oil sand, and permanent reservoir monitoring (see Section 2.4).

There are several examples of the time-lapse study in EOR and CCS, for example, Weyburn-Midale in Canada, Sleipner in offshore of Norway, In Salah in Algeria, RCSP-Phase II and III in the United States, Nagaoka in Japan, CO_2-CRC Otway in Australia, and Ketzin in Germany. More details are given in Chapter 6.

In 2007, CO_2 injection experiment was carried out in Nagaoka in Japan. The size of CO_2 injection was ca. 10,000 t in total; it was not as large as those in Weyburn-Midale in Canada, Sleipner and Snøhvit in Norway, and In Salah in Algeria. The progress of CO_2 injection was detected by loggings and cross-hole tomography. The sonic loggings were carried out and the change of V_P between before-injection and post-injection was 28% in the 5 m thick layer. The travel-time inversion was also carried out as a part of cross-hole seismic tomography studies. In contrast to travel-time inversion, waveform inversion using the same dataset gave 17% velocity changes. The big difference between two seismic approaches is important to know. The next stage of CO_2 injection has been carried out in offshore Tomakomai, Japan. Details of the Nagaoka experiment are given in Sections 2.3 and 2.4.

The CO_2 CRC-Otway program is in Australia. The program in Ketzin in Germany, is another scientific CO_2 injection experiment. The total amount of injected CO_2 in Ketzin between 2007 and 2013 was approximately 62,271 t. Geophysical and chemical measurements have been carried out to learn how CO_2 stays in the initial storage zone and the time lapse of the storage zone (Ivanova et al., 2012). Kasahara et al. (2013a) carried out a simulation assuming a seismic ACROSS source with some geophone arrays. Additional details of the Ketzin experiment and the result of simulation are given in Section 4.3.

The CO_2-EOR in Weyburn-Midale in Saskatchewan, Canada, was carried out from 2000 to 2011; 20 Mt of CO_2 in total was injected to subsurface. The time-lapse studies showed the clear image of reflection amplitude changes (Whitttake and Wildgust, 2011). Other time-lapse studies have been started in Aquistore in Saskatchewan, Canada, starting in 2014 (White et al., 2014). The 4D seismic, passive seismic, ACROSS time-lapse and the time-lapse studies using downhole measurements are ongoing. More details are given in Chapter 6.

Japan Canada Oil Sands Limited (JACOS) is doing relatively small-scale oil production from the heavy-oil reservoirs contained in the oil-sand layers. The depth of oil-sand reservoir is very shallow, such as 300 m depth. This is one of the examples of EOR efforts for the heavy oil production. The steam-assisted gravity drainage (SAGD) method has been used in this field. In the United States, the thermal EOR including steam and gas EOR including miscible and immiscible EOR are 300,762 b/d and 308,564 b/d in 2012, respectively (Koottungal, 2012). Four-dimensional seismic observations in the JACOS field between 2002 and 2006 were carried out to find the change of locations and physical property changes. The change was clearly revealed by the 4D seismic approach. Assuming this shallow oil-sand layers as the heavy oil reservoir, a simulation was carried out to determine whether the monitoring method using single source and multireceivers is feasible or not. This is described in Section 4.4.

1.5 FACTORS AFFECTING THE TIME-LAPSE STUDY

There are several factors affecting the repeatability of the time-lapse measurements (Table 1.1). However, the repeatability could be controlled by the system during the measurements and the temporal variation of the reservoirs themselves. If the seismic source is placed at fixed location, the some problem can be reduced. This is the case of ACROSS seismic source. If geophones are placed at some depth such as 50 m, the near-surface effect can be reduced. However, even if geophones are at downhole, the later phases could be affected by factors at near-surface layer as seen in Chapter 8.

1.6 SUMMARY OF TIME-LAPSE APPROACHES

Time-lapse approach has been frequently used in EOR in oil and gas exploration/production. There are many approaches to the time lapse, including 4D seismic method, 4D electromagnetic (EM) method, sonic, resistivity and neutron porosity loggings in downhole, cross-well seismic or EM tomography, vertical seismic profile (VSP), chemical sampling, ACROSS time-lapse method, etc. In this book, we will mainly describe the ACROSS seismic time-lapse method.

The time-lapse data are obtained by residuals among measurement in plural times. In 4D seismic reflection study, prestack or poststack seismic dataset are used. In the 4D reflection study, the acquisition circumstances are the same in both measurements or appropriate correction should be applied. This is the issue of repeatability and it is

Table 1.1 Factors affecting the time-lapse measurements

		Onshore	Offshore	Correction	Control
Source signature	Time base	*****	*****		Needed
	Ground coupling	*****			
	Gun pressure		*****	Needed	Needed
	EM bulb relay		*****	Needed	Needed
	Gun towing depth		*****		Needed
	Gun towing condition		*****		Needed
Structural change between source and geophone		Unknown	Unknown		
Ground roll		*****	*	Evaluate	
Surface (multiple) reflections		***		Evaluate	
Near-surface effect	Precipitation	*****		Evaluate	Measurement
	Temperature	*****		Evaluate	Measurement
	Seasonal	**			
Water column		None			
	Ocean tide level		*****	Needed	
	Salinity		*****	Needed	
	Temperature		*****	Needed	
	Bottom current		***		
	Ocean waves		***		
Aquifer level		*****			

Geophone(s)	Ground coupling	*****		Cementing	
	Streamer towing depth		*****		Control
	Streamer towing direction		*****		
	OBC/OBS position error		***	Needed	
	Response	***		Calibration	
	Drift	**	**	Calibration	
	Time base	***			Needed
	Tube wave in borehole	***		Estimate	
Positioning	GPS		*****		Control
	Navigation		*****		
Environmental noise	Traffic	*****		Stacking	
	Human activities	*****		Stacking	
	Weather	**	***		
	Snow	**			
Equipment corrections		*****	*****	Needed	Needed

The number of asterisks (*) indicates the degree of each effect.

quite a tough problem. In the cross-hole seismic tomography and zero-offset or walkaway VSP, the same data-acquisition systems or necessary correction are needed. If the source location(s) are fixed, a large part of problem in repeatability can be solved. In the ACROSS seismic time-lapse method, the ACROSS seismic source is mounted in a heavy concrete baseplate, which eliminates the major difficulties in repeatability.

CHAPTER 2

Various Time-Lapse Methods

Contents

2.1 4D SEISMIC METHOD

There are various time-lapse methods. The 4D seismic method that is most popular is the extension of 3D seismic survey. Offshore, 3D seismic survey is standard exploration. The data acquisition is easier than onshore. However, because the cost of one 3D survey is expensive, the time interval between two 3D surveys might be longer than a half year at best. In addition to the survey costs, it is necessary to do numerous corrections for the time lapse. The repeatability of the time lapse for offshore might be controlled by positioning of sources and receivers. In permanent reservoir monitoring (PRM) using ocean bottom cable (OBC) system, extensive time-lapse studies have been carried out in the North Sea, Azerbaijan, Gulf of Mexico, and offshore of Brazil. Using the OBC system, because the geophones and hydrophones are laid down on the sea floor, the correction on receiver locations and coupling to the sea floor could be less problematic. For the source side, the (air gun's) source signatures and the water column corrections should be solved. The overburden problems due to shooting positioning might cause repeatability errors. The surface seismic survey using the multi-streamers might require careful treatment for the correction for the repeatability.

Time Lapse Approach to Monitoring Oil, Gas, and CO_2 Storage by Seismic Methods
ISBN 978-0-12-803588-7
http://dx.doi.org/10.1016/B978-0-12-803588-7.00002-9
Copyright © 2017 Elsevier Inc.
All rights reserved.

Onshore, the misallocation of positions of sources and receivers causes more than 10% of repeatability in RMS errors. If people want to get better repeatability on shore, source(s) should be placed at fixed position and receivers should be buried at several tens of meters. Even if these conditions are satisfied, Jervis et al. (2012) think there are some inevitable errors.

2.2 CROSS-HOLE SEISMIC TOMOGRAPHY AND VERTICAL SEISMIC PROFILE

The surveys using wells are used for the time-lapse monitoring. Cross-well seismic tomography was used in the Nagaoka carbon capture and storage (CCS) experiment during the injection of CO_2 into the ground (see Section 6.1). The time-lapse estimation by travel-time tomography method tends to underestimate for the size of changing zone and magnitude for the temporal variation because the ray paths having the velocity decrease could be detoured as the rays running through the unchanged zone come first.

In the cross-hole seismic tomography, downhole seismic sources are used. The piezoelectric seismic source has less power than land vibrators, and the frequency given by the piezoelectric source is higher than land vibrators, weight drops, and air guns. If the frequency is 1 kHz, the wavelength is approximately a meter order to cause apparent attenuation by scattering. The maximum distance capability by piezoelectric source could be 500 m maximum. For the time-lapse study using piezoelectric seismic source, plural observation wells should be drilled very close to estimated temporal changing zones.

For the downhole sensors, distributed fiber sensor (DFS) such as distributed acoustic sensor (DAS) and distributed temperature sensor (DTS) can be extensively used in the downhole sensors for the time-lapse study.

For the case of vertical seismic profile (VSP), the combination in which seismic sensors are in well(s) and seismic source is at the ocean surface or on the ground surface is easier to do. Walk-away VSP has been used in many time-lapse cases.

2.3 WELL LOGGINGS

Repeated loggings such as sonic, dipole sonic, seismic, and resistivity have been used in many time-lapse studies. Measured values of several trips are compared. This method could be useful if the temporal changing zone is close to the logging wells. However, the interpretation of observed logging data is the issue whether they are heterogeneity or real-temporal change.

2.4 OCEAN BOTTOM CABLE/OCEAN BOTTOM SEISMOMETER FOR PERMANENT RESERVOIR MONITORING

Offshore, OBC has been used for PRM. OBCs with 4000–6000 channels have been used in the North Sea, Azerbaijan in the Caspian Sea, and offshore Brazil. Conoco Phillips Co. ordered 4D monitoring systems from six manufacturers for the Ekofisk oil field (Ekofisk Life of Field Seismic project) (Folstad, 2011). Optoplan AS Co. deployed 3966 ch. 4C OBS in 2010 and it covers 60 km^2 (Nakstad and Langhammer, 2011). BP deployed OBC so-called Life of Field Seismic system to Clair oil field (De Jongh, 2011; Ricketts and Barkved, 2011) and reported the temporal changes on saturation, pressure, and low-permeability matrix. Valhall oil field at 125 km southwest of Stavanger, Norway, is a soft chalk reservoir in the sea. The reservoir depth is 2500 m below 70 m water depth. In Valhall oil field, cable network of OYO Geospace 4C OBC 120 km long covers 45 km^2. Valhall has the world's first permanent 4D seismic array, and distinct temporal changes have been detected (Barkved, 2011).

Azerbaijan is flanked by the Caspian Sea. The first oil wells were drilled in 1847. The Azeri-Chirag-Gunashli (ACG) oil field is the world's largest producing offshore field, producing oil at 900,000 bbl/day rate. BP has carried out Chirag Azeri Reservoir Surveillance Project (CARSP) in ACG (Robinson et al., 2011; Seaborne et al., 2011). In this oil field, 4D surveys were done in 2007, 2008, and 2010 using 22 lines of 4C OBC array 6 km long (Seaborne et al., 2011). The sensor spacing of OBC is 75 m for 4 km long at the center and 150 m for the outer 2 km long. The distance between two OBCs is 500 m. Using 4D processing of the 2007 and 2008 CARSP data, water movement was detected on the north flank of Chirag and more significantly in the Central Azeri region (Robinson et al., 2011).

2.5 INTERFEROMETRIC SYNTHETIC APERTURE RADAR AND SEISMIC INTERFEROMETRY

Interferometry is another tool for time-lapse methods. Interferometric synthetic aperture radar (InSAR) is the interferometry method using satellite data. In the CCS of In Salah, Algeria, surface deformation was revealed by InSAR analysis (Vasco et al., 2008, 2010; Mathieson et al., 2010). The resolution given by InSAR is mm order and the ground deformation caused by In Salah injection was found as 20 mm uplift around KB-5 well. Although the storage zone of CO_2 is 1900 m in depth for In Salah, the amount of injected CO_2 was 3.8 Mt since 2004 (Ringrose et al., 2009, 2013) and the deformation around the wells (KB-5, KB-502 and KB-503) was measured. The 4D seismic measurement in 2009 showed slight depression along the linear trends near the KB-502 and KB-503 wells (Ringrose et al., 2009, 2013).

In contrast to the interferometry using satellite radar, the seismic interferometry can be used for the time-lapse problem. Our approach of seismic interferometry using traffic noise showed similar temporal changes revealed by the seismic active source (ACROSS) (see Section 5.3).

2.6 DISTRIBUTED TEMPERATURE SENSOR

The DTS can be used for the time-lapse study in the downhole measurement and at the ground surface. In the case of the steam enhanced oil recovery at deep reservoirs, the surface DTS measurement could not be enough to detect the temperature change due to the steam injection.

CHAPTER 3

Active Seismic Approach by Accurately Controlled and Routinely Operated Signal System

Contents

3.1 UNIQUENESS OF THE ACCURATELY CONTROLLED AND ROUTINELY OPERATED SIGNAL SYSTEM (ACROSS) APPROACH

3.1.1 Introduction

We treat the description of some objects, such as oil and gas reservoirs, in 3D and/or 4D space. In most cases, the description of the heterogeneities of the physical properties of objects in 3D space is sufficient. Oil and gas exploration requires accurate description of reservoir characteristics in 3D space. However, after geophysical reservoir mapping, the temporal change in physical and chemical states of any object must be monitored. In such cases, we use physical

Time Lapse Approach to Monitoring Oil, Gas, and CO_2 Storage by Seismic Methods
ISBN 978-0-12-803588-7
http://dx.doi.org/10.1016/B978-0-12-803588-7.00003-0
Copyright © 2017 Elsevier Inc.
All rights reserved.

and chemical parameters to determine the change in physical and chemical states from time to time.

In geophysical exploration, we frequently use seismic waves and/or electromagnetic waves to obtain the measurements. In this book, we discuss the temporal change of seismic waves used to monitor the physical state of the subsurface.

If some physical properties, such as velocities, reflection coefficients, attenuations, and density in the subsurface, are changed with time by injection of steam or supercritical CO_2, fracking of the subsurface, or just conventional oil and gas production, the change in the physical properties may modify the propagation paths of seismic waves or create diffractions (new scattering fields). The Accurately Controlled Routinely Operated Signal System (ACROSS) can be used for nearly continuous monitoring of temporal change in seismic wave propagation.

Reflection seismology uses the seismic reflectivity distribution of the subsurface in 3D and/or 4D space. Understanding the temporal change of the subsurface first requires temporal-change data of the 3D signature of the subsurface, that is, a 4D seismic survey. It is simply obtained by multiple 3D surveys. However, some major discrepancies exist between 3D and 4D seismic surveys.

Because seismic waves are frequently used in geophysical exploration of oil and gas resources, we introduce the uniqueness of the ACROSS approach in seismic surveys.

Although it is not frequently used in exploration seismology, a description of the "transfer function" or "Green's function" as shown in Fig. 3.1 is necessary. For simplicity, we assume the single force $F_i(t)$ applies to the ground at the source in direction i, and the displacement $U_j(t)$ is observed at the receiving point in direction j, where each i and j is either of x, y, or z. The transfer function between two locations $H_{i,j}(t)$ is given in Fig. 3.1 and Eq. (3.1).

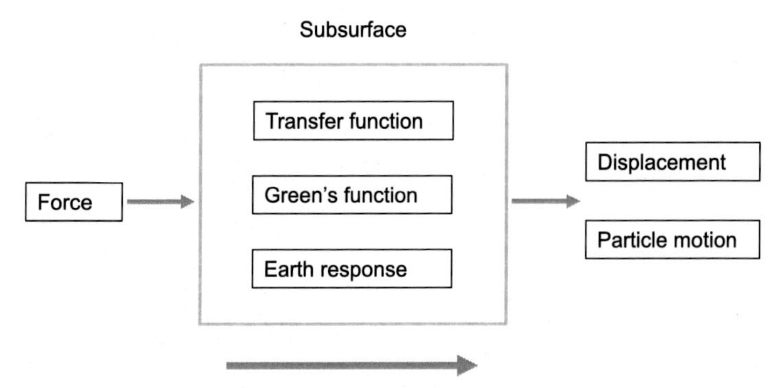

Figure 3.1 Definition of "transfer function" and "Green's function." If input is force and output is displacement, the transfer function corresponds to Green's function in elastic wave theory. It is also called the earth response.

The relations among U_j, $H_{i,j}$, and F_i are

$$U_j(t) = H_{i,j}(t) * F_i(t). \tag{3.1}$$

F_i and U_j are vectors and $H_{i,j}$ is a second-rank tensor. The asterisk denotes a convolution operator.

As described later, the source signature will be changed as the ground-coupling condition caused by placement of the source on the ground, even if the location of the source is the same and might depend on rainfall or characteristics of the ground layers. At the ocean, source signatures of an air gun array may be changed by sea states, towing methods, air gun pressure, and time delays of the electromagnetic bulb switch of each gun.

For a simple case in the frequency domain

$$U_j(\omega) = H_{i,j}(\omega) \cdot F_i(\omega), \tag{3.2}$$

where ω is angular frequency.

$H_{i,j}$ is described by the physical properties—velocities, attenuations, and density—between the source and the receiving point if these physical properties are independent of the frequency.

For two-dimensional cases, forces are vertical and horizontal, and displacements are also vertical and horizontal.

3.1.2 Cross-Correlation of Observed Records and Source Signature

In the processing of the ordinal seismic survey, cross-correlation of observation U and source F in the time domain is applied as

$$U_j(t) * F_i(-t) = H_{i,j}(t) * F_i(t) * F_i(-t). \tag{3.3}$$

Eq. (3.3) equals the convolution of transfer function $H_{i,j}(t)$ and the autocorrelation of the source $F_i(t) * F_i(-t)$, which is a zero-phase wavelet.

In the frequency domain,

$$U_j(\omega) \cdot \overline{F_i(\omega)} = H_{i,j}(\omega) \cdot F_i(\omega) \cdot \overline{F_i(\omega)} = H_{i,j}(\omega) \cdot |F_i(\omega)|^2, \tag{3.4}$$

where $|F_i(\omega)|^2$ is the power spectrum of the source, and $\overline{F(\omega)}$ is the complex conjugate of $F(\omega)$.

We can obtain $H(t) * F(t) * F(-t)$ or $H(\omega) \cdot F(\omega) \cdot \overline{F(\omega)}$.

The autocorrelations of force $F(t) * F(-t)$ for two sweep patterns are shown in Fig. 3.2 as examples. If the ratio of the up and down sweep is 1:1, the shape of the sweep is triangular. The autocorrelation of the triangular sweep is shown for the sweep from 10 to 300 Hz during 10 s. If the source sweep is kept as a constant force (constant acceleration), autocorrelation of the source, $F(t) * F(-t)$, is close to a zero-phase impulse with side lobes. The width of the side lobes is ~0.05 s because of the wider frequency bands such as 10–300 Hz. In the case of a saw-tooth sweep of 15:1 for the up and down sweep, the autocorrelation shows similar side lobes, but ringing of the side lobes is smaller than around the triangular sweep. These side lobes might disturb the resolution of reflection records if an interpreter evaluates the thickness of each layer. The deconvolution process is an attempt to minimize the effects of the side lobes and get the sharpest possible impulse. For most seismic reflection surveys, this method is acceptable.

3.1.3 4D Seismic Survey

In repeated 3D surveys, the so-called 4D seismic survey is

$$U_j(\omega, T_a) = H_{i,j}(\omega, T_a) \cdot F_i(\omega), \tag{3.5}$$

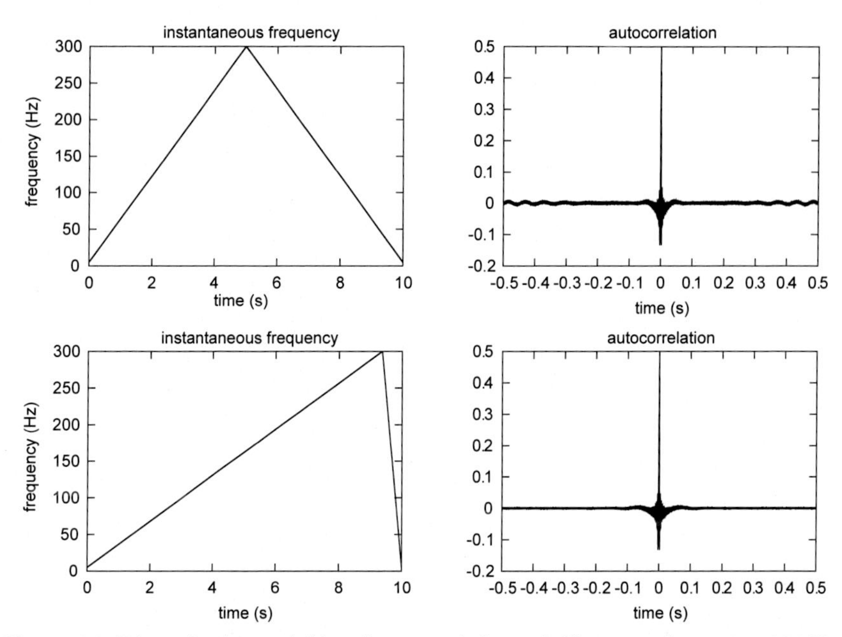

Figure 3.2 Triangular (upper left) and saw-tooth (lower left) sweeps from 10 to 300 Hz, with a constant force during 10 s. The autocorrelations of two kinds of sweeps are shown in the right panels. Both autocorrelations show side lobes beside the zero-phase impulse.

where $U_j(\omega, T_a)$ is the displacement measurement in direction j at time T_a, and $F_i(\omega, T_a)$ is the source signature in direction i at time T_a.

If the source has the same signature at time T_a and time T_b,

$$U_j(\omega, T_b) = H_{i,j}(\omega, T_b) \cdot F_i(\omega) \tag{3.6}$$

$$\delta U_j(\omega, T_a, T_b) = U_j(\omega, T_a) - U_j(\omega, T_b) = \delta H_{i,j}(\omega, T_a, T_b) \cdot F_i(\omega) \tag{3.7}$$

where $U_j(\omega, T_a)$ and $U_j(\omega, T_b)$ are subsurface responses at times T_a and T_b.

When we make a cross-spectrum with source $F_i(\omega)$,

$$\begin{aligned}\delta U_j(\omega, T_a, T_b) \cdot \overline{F_i(\omega)} &= (H_{i,j}(\omega, T_a) - H_{i,j}(\omega, T_b)) \cdot F_i(\omega) \cdot \overline{F_i(\omega)} \\ &= \delta H_{i,j}(\omega, T_a, T_b) \cdot |F_i(\omega)|^2,\end{aligned} \tag{3.8}$$

we obtain $\delta H_{i,j}(\omega, T_a, T_b) \cdot |F_i(\omega)|^2$, which is the multiplication of the power spectrum of the source signature and the residual transfer function for the subsurface characteristics that describe how the subsurface changes and the location of the changes.

Using Eq. (3.8) for many source–receiver pairs, we can map the physical properties distribution of the subsurface. These distributions could be related to oil and gas reservoir changes and the physical status of carbon capture and storage.

3.1.4 Factors Affecting the 4D Seismic Survey

Eqs. (3.5) and (3.6) assume the source signature is the same between time T_a and time T_b, or we can precisely match the source signatures for two surveys.

In an ordinary seismic survey, the source signature $F_i(t)$ or $F_i(\omega)$ is controlled as much as possible or the source signature is compensated to be the same for two observations. However, the coupling of the source vibrator on the ground and the near-surface properties just beneath the source may change from place to place. This change might modify the source signature, including in the area surrounding the source. Even if the operator of the vibrator tries to maintain the amplitudes of the source at the same level as whole surveys, the source signatures may vary from one survey to another. This variation might cause ambiguity in time-lapse resolution.

At the ocean, the source signature of a tuned air gun may be affected by gun pressure, delay times of each air gun, depth of the air gun array, and towing speeds. The tidal force may change the sea level from time to time.

As described as earlier, the time lapse always presents questions about the similarity of source and receiver characteristics among plural surveys.

3.1.5 The ACROSS Approach

The ACROSS methodology was proposed by Kumazawa and Takei (1994), just before the Great Hanshin earthquake around Kobe City ($Mw = 6.9$). The concept of ACROSS has been extended to the other fields such as electromagnetic approach, so that the ACROSS method using seismic wave is referred as the seismic ACROSS.

These researchers built two ACROSS seismic systems, one at the Tono Geoscience Center of the Japan Atomic Energy Agency (Kumazawa et al., 2000; Kunitomo and Kumazawa, 2004) and the other on Awaji Island (Yamaoka et al., 2001). The system at Tono had been operated nearly continuously for several years. The main task of the seismic-ACROSS approach is to minimize the temporal variation of the source system. To minimize the temporal variation of the source signature, the source should be fixed on a relatively heavy concrete block, for example, 5 m long × 5 m wide × 3 m thick. This method of installation will restrict the mobility of the source. To demonstrate how to overcome this weakness, we performed simulations assuming a few sources and a number of receivers to retrieve the diffracted location caused by injection of CO_2 or steam, or fracking (see Chapter 4).

Two kinds of seismic sources are used for the ACROSS methods: rotary-type and linear-motion-type. For a rotary-type seismic source, the rotation of an eccentric mass is controlled by a servomotor. The mass position is precisely controlled typically at 8192 counts/cycle.

In the ACROSS method, a commonly used source signature is a sweep from f_a to f_b during sweep duration time T_m. The sweep comprises up and down sweeps. In our case, we use the ratio up:down = 7:1 or 15:1 to get smooth amplitude spectra. If we use a triangular sweep with up:down ratio of 1:1, the amplitude spectra show strong amplitude oscillation with frequency. This oscillation causes a low signal-to-noise ratio (S/N) at the frequencies with small source amplitudes.

Use of a frequency sweep with time (chirp signal) is similar to a conventional land vibrator, but the method of analysis is different from those commonly used in exploration seismology. In the processing of Vibroseis data, the cross-correlation of the source signature and the observed seismogram is used as described in Section 3.1.2. This cross-correlation generates convolution of the autocorrelation of the source signature and the subsurface transfer function. To get the subsurface transfer function, a deconvolution process such as predictive deconvolution, spiking deconvolution, or minimum-phase filter must be applied to improve the resolution. On the other hand, the ACROSS method uses source-signature deconvolution in the frequency domain.

The following are characteristics of the rotary-type ACROSS seismic source:

a) Line spectra: Frequency is precisely controlled by the GPS time base. By use of the GPS time base, 10-μs accuracy can be achieved. When a T_m-long dataset is transformed to the frequency domain, the spectrum of this dataset is a set of line spectra with $\Delta f = 1/T_m$ frequency spacing. Details are described in Sections 3.2 and 3.3.

b) Force: For a rotational source, as seen in Fig. 3.3, the force generated by the rotation of the weight M of the eccentric mass is

$$F = MR\omega^2, \tag{3.9}$$

where ω and R are the angular frequency and the radius of rotation of mass, respectively.

c) Simultaneous vertical and horizontal vibrations: By combining clockwise and counterclockwise rotations, we can synthesize any directional vibrations perpendicular to the rotation axis. The vertical and horizontal vibrations are vertical and horizontal single forces, respectively. The clockwise and counterclockwise rotations are alternatively operated for 1 h each. The details are described in Appendix B.6. For the linear-motion-type ACROSS seismic

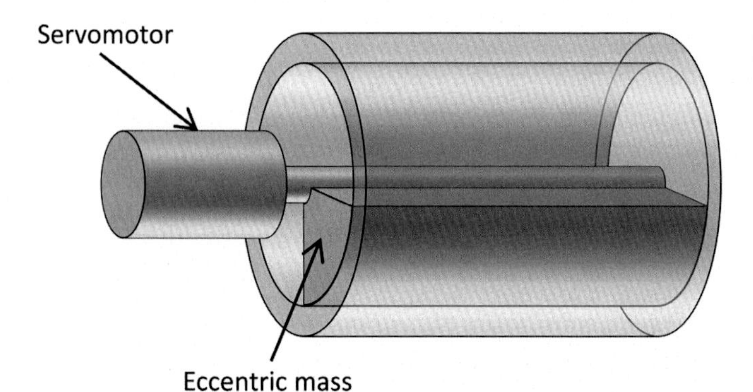

Figure 3.3 Principle of the rotary-type ACROSS seismic source as used in a horizontal rotation axis case.

source this procedure cannot be applied. In the ACROSS study, the P and S were clearly separated by this processing (Kasahara et al., 2011b).

d) Separation of vibrational signal and noise spectra, and weighted stacking: If the $2T_m$-long dataset is transformed to the frequency domain, the spectra comprise signal and noise line spectra, where $2T_m$ is the basic dataset time frame. The spectral spacing is $1/(2T_m)$. The signal and noise distribute on even and odd frequencies, respectively. By this processing, we can quantitatively evaluate the noise spectra and S/N ratio. We use the S/N ratio for the weighted stacking process described in Section 3.3 (Fig. 3.12).

e) Simultaneous excitation of plural sources: It is possible to operate plural sources simultaneously and receive their signals independently by increasing data length and shifting the career frequency of each.

f) Separation of natural earthquakes during active source operation: By this method, it is possible to separate an active-source signal series from natural noise, natural earthquakes, and fracking events.

g) Seismic interferometry using background noise: If we calculate the cross-spectrum of noise at receivers A and B, we can obtain cross-correlation for seismic interferometry of stations A and B, even when the ACROSS source is in operation. The noise could be generated naturally and/or by human activity, such as automobile traffic or industrial processes. This application is described in Section 5.2.

h) Enhancement of S/N: To get better S/N, the stacking of observed data applies to either the frequency domain or the time domain. Because both the source control and the data sampling are based on the identical accurate time standard, we can stack the dataset only on each designed spectral line. By the stacking, the S/N can be improved by $\sqrt{n}$, where n is number of stacks.

i) Transfer function in the frequency domain: In seismic ACROSS processing, observed seismograms in the frequency domain are divided by source signatures in the frequency domain. Each of the results is the subsurface response or the transfer function between source and receiver in the frequency domain.

j) Transfer function in the time domain: The subsurface response or the transfer function in the frequency domain is transformed to the time domain. However, because the frequency band of the sweep is limited, the band-limited frequency-window effect remains, causing side lobes on a zero-phase wavelet (see Section 3.3.4). We use a cosine-taper window in the frequency domain when we transform the transfer function from the frequency domain to the time domain. Then we can obtain time domain seismograms for the force with a frequency band from f_a to f_b.

k) Description of the second-rank tensor of the transfer function: When we describe the transfer function, we use the following expressions:

H_{zV} is the vertical force observed by the UD-component geophone.

H_{xH} or H_{yH} is the horizontal force observed by the NS- or EW-component geophone.

The two suffixes show these expressions are the second-rank tensor. The first and second suffixes denote directions of the receiver and the source, respectively.

The methods of ACROSS phase control and the switching between clockwise and counterclockwise rotations are described in Appendices B.5 and B.6, respectively.

Table 3.1 is the summary of the ACROSS seismic source and its processing. The details are described in Section 3.3 and Appendix C.

3.2 OUTLINE OF THE ACROSS SEISMIC SOURCE SYSTEM (SEE APPENDIX B FOR DETAILS)

This section describes the outline of a rotary-type ACROSS seismic source (Figs. 3.4 and 3.5).

The ACROSS seismic source is comprised of the following:

(A) Eccentric mass

(B) Servomotor (Fig. 3.4)

(C) Motor control unit

Table 3.1 Comparison of vibroseis and the rotary-type ACROSS seismic source

Seismic source	Vibroseis	Rotary-type ACROSS seismic source
	Frequency sweep	Frequency sweep controlled by GPS time base: (a)
Vibration direction	Vertical or horizontal vibration	Synthesize: (c)
Source control	GPS	GPS time base: (a)
	Flat acceleration	Linear or nonlinear sweep
	Feedback	Mass position controlled by the servomotor
		Clockwise or counterclockwise rotation
Processing	Cross-correlation	Discrete Fourier transform
		Separation of signal and noise using line spectra of designed sweep function: (d)
	Stacking in time domain	Stacking of received signals in spectral domain: (h)
		Received signal divided by source signature gives transfer function: (i)
		Calculation of vertical and horizontal vibration response
	Deconvolution	Inverse Fourier transform: (j)
		Noise interferometry: (g)

(a)−(g) are described in the previous text.

Figure 3.4 Rotary-type ACROSS seismic source used in Saudi Arabia.

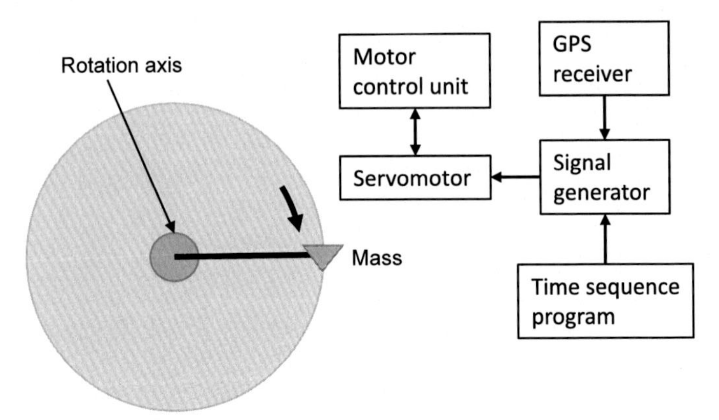

Figure 3.5 Summary of the rotary-type ACROSS seismic source system.

(D) Signal generator (Fig. 3.6)
(E) GPS time base
(F) System control PC

The main part of the ACROSS seismic source is a servomotor (B and Fig. 3.4). The motor rotates an eccentric mass (A) (Fig. 3.5). The position of the mass is controlled by a servosystem consisting of the motor control unit (C), with a typical accuracy of 8192 counts per one rotation. The signal generator (D) (Fig. 3.6) generates the number of pulses corresponding to the designed function of the mass position and supplies the motor-control unit with the GPS time base (E). The control screen of the system-control PC (F) is shown in Fig. 3.7.

Figure 3.6 Signal generator for the rotary-type ACROSS seismic source.

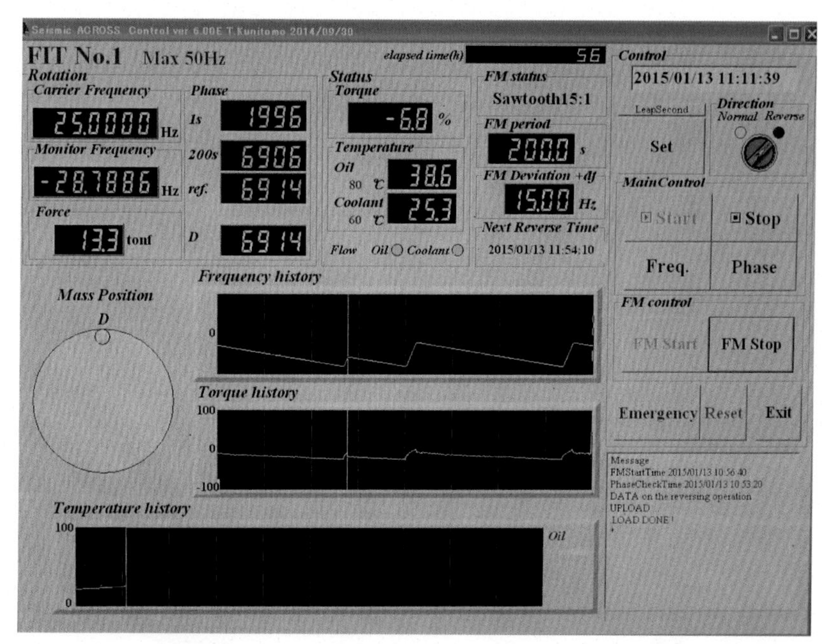

Figure 3.7 Control screen of the rotary-type ACROSS seismic source system.

Because the force generated by a rotational source is given by Eq. (3.9), low-frequency and high-frequency forces are smaller and larger, respectively. The longer duration in the low-frequency sweep and shorter duration in the high-frequency sweep are used in the relative sense. The frequency-sweep type in which the instantaneous frequency is a quadratic or higher-degree function of time is called a nonlinear sweep. Fig. 3.8 shows the pattern of nonlinear sweep from 10 to 50 Hz during 200 s. In contrast, if the instantaneous frequency is a linear function of time, it is called a linear sweep. By use of a nonlinear sweep, the frequency dependency of the force can be improved. The comparison of linear and nonlinear sweeps is shown in Fig. 3.9. The up:down ratios of 15:1 and 1:1 are also shown in Fig. 3.9.

Observed near-field ground motions adjacent to the source are different for each component (Fig. 3.10). The ground motions on the vertical (V) and horizontal, perpendicular to the rotational axis (H1), are at higher levels than H2. The amplitude of the observed

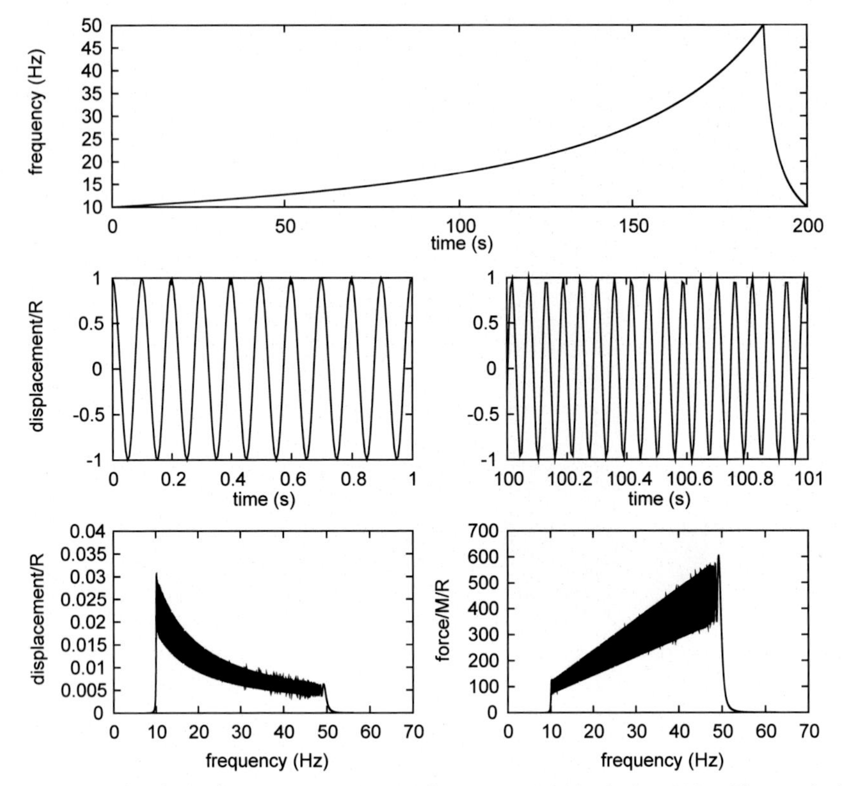

Figure 3.8 (Top) Nonlinear sweep pattern from 10 to 50 Hz during 200 s. The vertical axis is frequency (Hz) and the horizontal axis is time (seconds). (Middle) Vertical displacements of the mass in parts of the above sweep between 0 and 1 s (middle left) and 100 and 101 s (middle right). (Lower left) Vertical displacement of mass/R vs frequency (Hz). (Lower right) (Vertical force)/MR vs. frequency (Hz), where *R* and *M* are rotation radius and mass weight, respectively.

ground motion at A1 and A4 are also different because of the radiation pattern.

Separation of the ACROSS vibration and background noise can be done as mentioned in Section 3.1.5. The method of separation of signal and noise is described in Section 3.3 and Appendices C.2 and C.3. Fig. 3.11 shows the observed near-field ground motion and the estimated noise levels. The S/N ratio is $\sim 10^3$ just beside the ACROSS seismic source.

When 400 s of the observed data are transformed to the frequency domain by discrete Fourier transform (DFT), the spectra of signal and

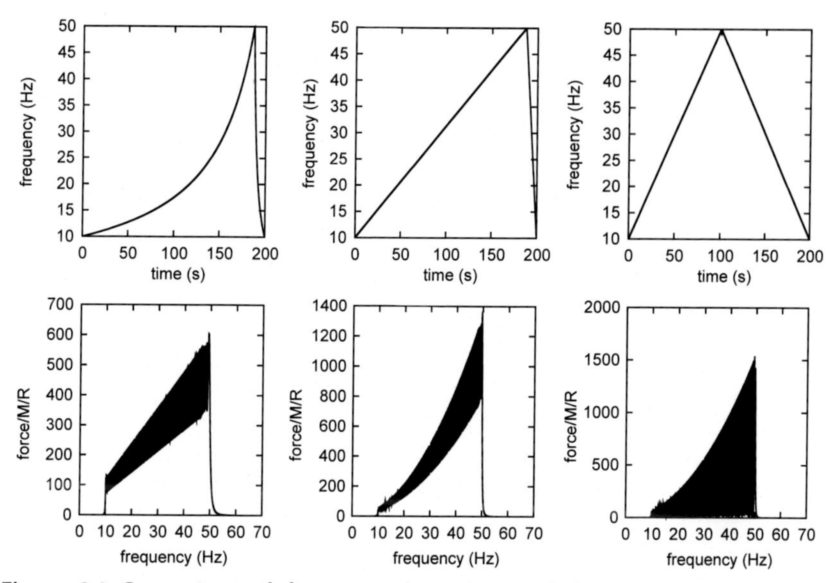

Figure 3.9 Comparison of frequency dependence of theoretical force for (left) nonlinear (up:down = 15:1), (middle) linear sweep (up:down = 15:1), and (right) linear sweep (up:down = 1:1) for the sweep of 10–50 Hz during 200 s. Upper figures: frequency (Hz) vs time (seconds). Lower figures: force/MR vs frequency (Hz), where R and M are rotation radius and mass weight, respectively.

noise are sets of line spectra. The signal spectra and noise levels estimated by noise spectra are shown in Fig. 3.11.

The generated signal and noise are distributed alternately.

Signal series: $10 + 0.005 * n$ ($n = 0$–3999) Hz.

Noise series: $10.0025 + 0.005 * n$ ($n = 0$–3999) Hz.

3.3 OUTLINE OF THE ACROSS DATA PROCESSING

3.3.1 Flow of Processing

The objective of seismic-ACROSS data processing is to obtain the transfer function between the source and the receiver. Fig. 3.12 shows the schematic flow diagram of processing for a rotary-type ACROSS seismic source. There are two starting points: the observed record and the source-movement data.

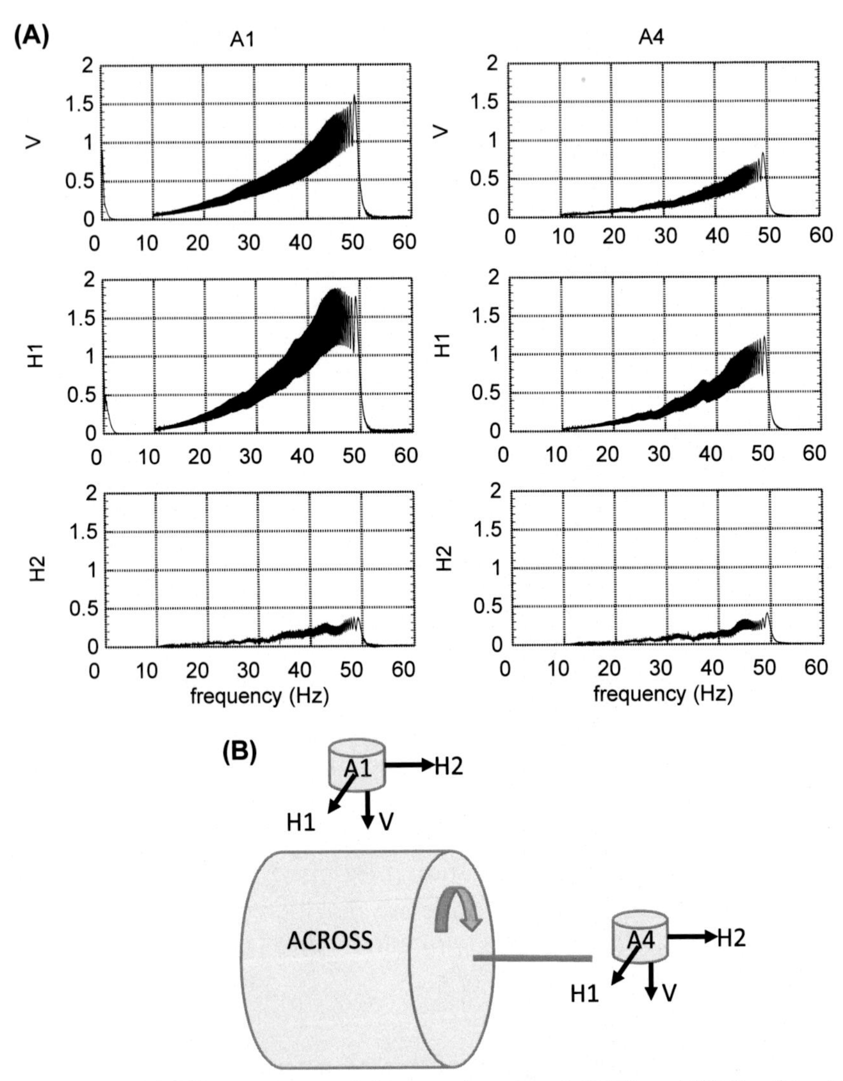

Figure 3.10 (A) Difference in near-field ground motions in V, H1, and H2 directions. V: vertical component. H1: horizontal component perpendicular to the rotational axis. H2: horizontal component parallel to the rotational axis. The sweep is nonlinear from 10 to 50 Hz during 200 s. The ratio of up and down sweeps is 15:1. Left column (A1): records of 3-component seismometer located in the radial direction of the rotational axis. Right column (A4): records of 3-component seismometer located in the axial direction of the rotational axis. (B) Layout of A1 and A4 seismometers.

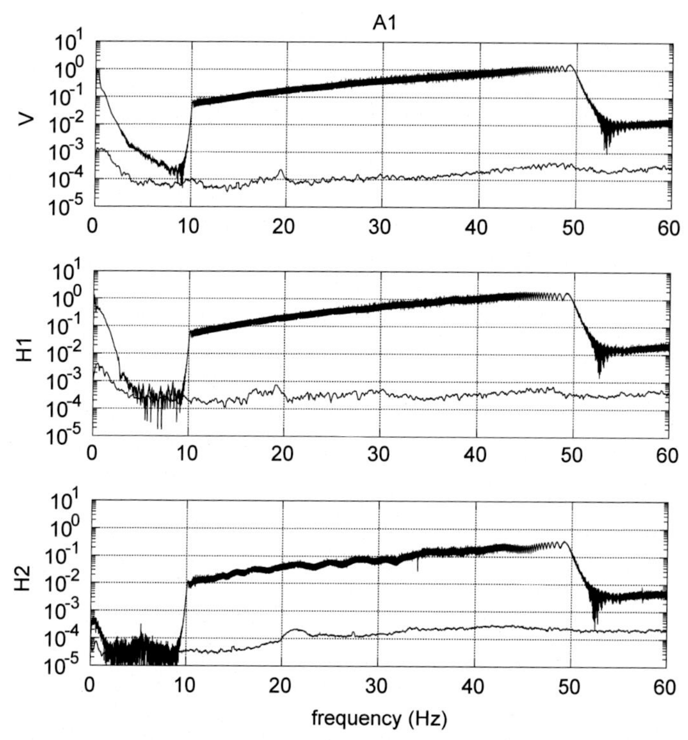

Figure 3.11 Observed near-field ground motion levels of V, H1, and H2 and the estimated noise levels observed near the ACROSS seismic source. S/N ratio is > 10^3.

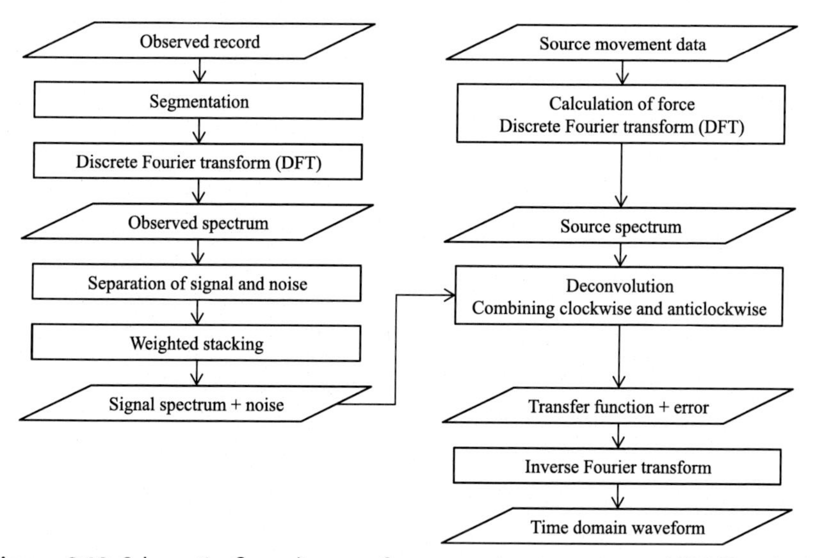

Figure 3.12 Schematic flow diagram for processing rotary-type ACROSS seismic source data.

First, the observed record at a receiver is divided into segments in accordance with the source operation, and then converted to the observed spectrum by DFT. One segment contains a certain number of sweep cycles from f_1 to f_2, for example, 200 s from 10 to 50 Hz. The source-signal components are extracted from the observed spectrum, and the noise levels are estimated.

On the other hand, the source spectrum is calculated from the source-movement data measured by the rotary encoder or obtained by the designed sweep function.

Subsequently, the observed signal spectrum is divided by the source spectrum to yield the transfer function in the frequency domain. The time-domain waveform is obtained by inverse Fourier transform, which we need in most cases.

Even for the other types of ACROSS seismic sources, the basic process is similar. Each process will be discussed in the following sections, and further details are given in Appendix C.

3.3.2 Processing the Observed Record at a Receiver

The observed record at a receiver is a time series of a certain length, typically 1 h. First, the record is segmented in accordance with the period of the source operation. When the repetition period of the source is T, the length of one segment in the time domain must be an integer multiple of T, such as mT.

The next process is application of the DFT. The DFT gives the complex values for integer multiple frequencies of $1/(mT)$. Unless the segment length is an integer multiple of the repetition period, the source signal will diffuse on the frequency axis. No tapers or zero-paddings are used in the DFT.

Spectral components for the other frequencies, except the integer multiple of $1/(mT)$, consist of background noises, which can be used to estimate errors in the signal spectrum. The source signal frequency is obtained by calculating the source spectrum as mentioned following, so that the signal spectrum can be extracted. The noise level at each signal frequency is calculated by the root mean squares of the neighboring noise frequencies.

3.3.3 Preparing the Source Spectrum

To obtain the transfer functions, we must prepare the "denominator" of a transfer function, which is the source spectrum. The source signal in an ACROSS observation is treated as known. For a rotary-type source, the movement of the mass can be known by the output of the rotary encoder. Otherwise, we can use the design function provided to the source controller if the controllability is nearly perfect.

If $\theta(t)$ is the time function of the mass position in the angle, the centrifugal forces $f_X(t)$ and $f_Y(t)$ in two orthogonal directions perpendicular to the motor axis are

$$f_X(t) = -MR\frac{d^2}{dt^2}\cos\theta(t), \quad f_Y(t) = -MR\frac{d^2}{dt^2}\sin\theta(t), \tag{3.10}$$

where M and R are the weight and radius of the eccentric mass, respectively.

In the frequency domain,

$$F_X(\omega) = MR\omega^2 C(\omega), \quad F_Y(\omega) = MR\omega^2 S(\omega), \tag{3.11}$$

where $C(\omega)$ and $S(\omega)$ are the Fourier transforms of $\cos\theta(t)$ and $\sin\theta(t)$, respectively.

The data length for the DFT must be the same as the repetition period of the source signal, in a similar manner to processing the observed record in the former subsection.

3.3.4 Calculating the Transfer Function

The transfer function $H(\omega)$, defined as Eq. (3.2), is calculated from the observation spectrum $U(\omega)$ and the source spectrum $F(\omega)$. If the two components of the source spectrum vector for clockwise rotation are defined as Eq. (3.11), those for counterclockwise are $F_X(\omega)$ and $-F_Y(\omega)$. By adding and subtracting the observation spectrum for clockwise rotation $U_j^{(+)}(\omega)$ and the one for counterclockwise rotation $U_j^{(-)}(\omega)$, synthetic observation spectra for two orthogonal forces are obtained.

$$\begin{aligned} U_j^{(+)}(\omega) + U_j^{(-)}(\omega) &= H_{jX}(\omega)F_X(\omega), \\ U_j^{(+)}(\omega) - U_j^{(-)}(\omega) &= H_{jY}(\omega)F_Y(\omega), \quad (j = x, y, z). \end{aligned} \tag{3.12}$$

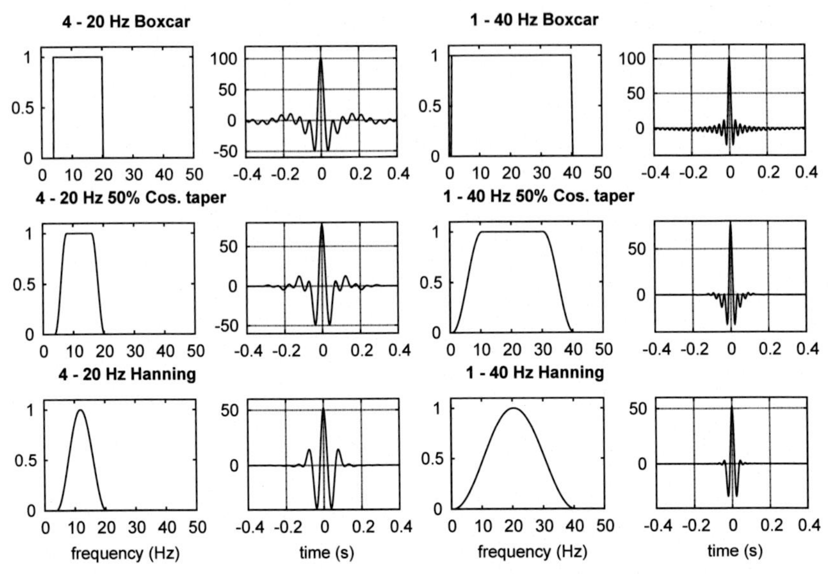

Figure 3.13 Effect of frequency windows on time domain waveforms. Each of the six examples is shown by frequency window (left) and wavelet pattern (right).

Therefore, six components of the tensor transfer function are calculated as

$$
\begin{aligned}
H_{jX}(\omega) &= \left(U_j^{(+)}(\omega) + U_j^{(-)}(\omega)\right) \Big/ F_X(\omega), \\
H_{jY}(\omega) &= \left(U_j^{(+)}(\omega) - U_j^{(-)}(\omega)\right) \Big/ F_Y(\omega), \quad (j = x, y, z).
\end{aligned} \tag{3.13}
$$

3.3.5 Time Domain Waveform

To discuss the seismic wave arrivals, the time domain waveform is useful. The simplest solution is the Fourier series,

$$
h(t) = \sum_{\omega} H(\omega)\exp(i\omega t). \tag{3.14}
$$

The sampling rate of time t must be considerably smaller than $2\pi/\max(\omega)$ for smoothness of the waveform. Another important factor is the effect of the frequency-range limitation. Application of an appropriate frequency window before transformation avoids side lobes on the wavelet of each arrival. Examples of frequency windows and their wavelet patterns in the time domain are shown in Fig. 3.13.

CHAPTER 4

Imaging of Temporal Changes by Backpropagation

Contents

In this chapter, we show imaging simulation of the time lapse for some cases.

4.1 BACKPROJECTION

Firstly, terminologies of backprojection and backpropagation are different. For example, travel-time inversion is backprojection (e.g., Claerbout, 1992) and waveform inversion is backpropagation.

We will briefly summarize the backprojection in case of seismic travel-time inversion as the case of shale-gas fracking quakes.

The observed data are travel times of plural stations for plural events, and the unknown parameters to be estimated are source locations, origin times of all events, and velocity structure.

Using observed travel times T_{ij}^{obs} at i-th station (x_i, y_i, z_i) $(i = 1, ..N)$ for j-th earthquake $j = 1, \ldots, M$ and theoretical travel times T_{ij}^{cal} under the initial model parameters (source

Time Lapse Approach to Monitoring Oil, Gas, and CO_2 Storage by Seismic Methods
ISBN 978-0-12-803588-7
http://dx.doi.org/10.1016/B978-0-12-803588-7.00004-2

Copyright © 2017
Elsevier Inc.
All rights reserved.

locations, origin times, and velocity structure), we will obtain the travel-time residuals,

$$\begin{aligned}\Delta T_{ij} &= T_{ij}^{obs} - T_{ij}^{cal} \\ &= \left(\frac{\partial T}{\partial x}\right)_{ij}\Delta x_j + \left(\frac{\partial T}{\partial y}\right)_{ij}\Delta y_j + \left(\frac{\partial T}{\partial z}\right)_{ij}\Delta z_j + \Delta t_j + \sum_k L_{ij}^k F_k + E_{ij},\end{aligned} \tag{4.1}$$

where $\Delta x_j, \Delta y_j, \Delta z_j$ and Δt_j are the correction amount of the source locations and origin times, F_k and L_{ij}^k are slowness perturbation and path length through k-th cell, and E_{ij} is the error.

Using homogeneous velocity model, theoretical travel time T_{ij}^{cal} is given by the following equation,

$$T_{ij}^{cal} = T_j^0 + \frac{\left\{\left(x_i - x_j^0\right)^2 + \left(y_i - y_j^0\right)^2 + \left(z_i - z_j^0\right)^2\right\}^{1/2}}{V_0}, \tag{4.2}$$

where $\left(x_j^0, y_j^0, z_j^0, T_j^0\right)$ is the first source parameters for the earthquake event number j, and V_0 is the initial velocity.

Observations are travel-time residuals,

$$Y = \Delta T_{ij} \tag{4.3}$$

Unknowns are $X = (\Delta x_1, \Delta y_1, \Delta z_1, \Delta t_1, \Delta x_2, \Delta y_2, \Delta z_2, \Delta t_2, \ldots. \Delta x_M, \Delta y_M, \Delta z_M, \Delta t_M, F_1, \ldots, F_k)$.

The equation between the observation Y and the unknown X is

$$AX = Y. \tag{4.4}$$

By the least square sense, we minimize the residual,

$$\min\|Y - AX\|. \tag{4.5}$$

The source parameters and slowness of the subsurface are obtained by solving this equation using singular value decomposition, conjugate gradient, LSQR (Paige and Saunders, 1982), and so on.

In the case of X-ray tomography, the attenuation of projected X-ray into the (human) body is measured to obtain the physical properties of (human) body. This is another case of projection. Radon transform is one kind of projection (e.g., Kasahara and Tomoda, 1993).

4.2 BACKPROPAGATION METHOD FOR THE TIME-LAPSE IMAGING

We assume single (or a few) seismic source(s) [e.g., Accurately Controlled and Routinely Operated Signal System (ACROSS) seismic source] and multigeophones as shown in Fig. 4.1. When a single force is applied on ground surface, it generates seismic waves and the waveform records are obtained by geophones at the surface and/or in boreholes. Seismic waves reflected or scattered at the target reservoir could be observed by the geophones.

The principle of the time-lapse imaging is shown in Fig. 4.2. The seismograms observed at distributed stations are time reversed and used as the source-time functions at the stations for wave-field synthesis called backpropagation. In the backpropagated wave field, seismic waves radiated from all the stations make a wave front together, and it finally converges to the original source point.

For time-lapse imaging using backpropagation, the residual waveforms calculated by subtraction of observed waveforms at two points of time are used. Because the residual waveforms only comprise scattered or diffracted waves from the changed zones, the backpropagated wave field converges to the changed zones, for example, target reservoir. In other words, the location causing the waveform change behaves as secondary seismic source.

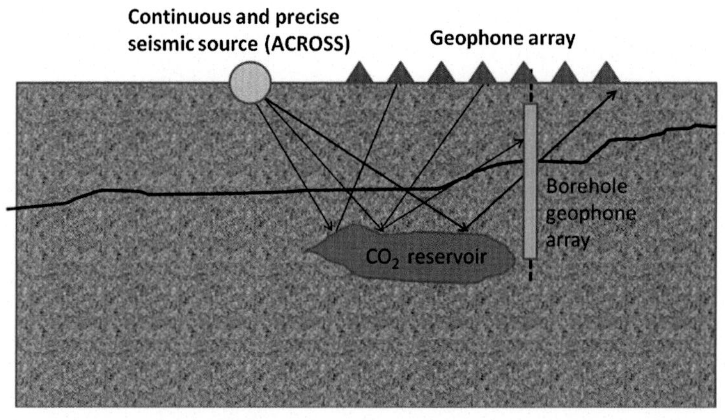

Figure 4.1 Concept of imaging by single seismic source (e.g., ACROSS seismic source) and a geophone array.

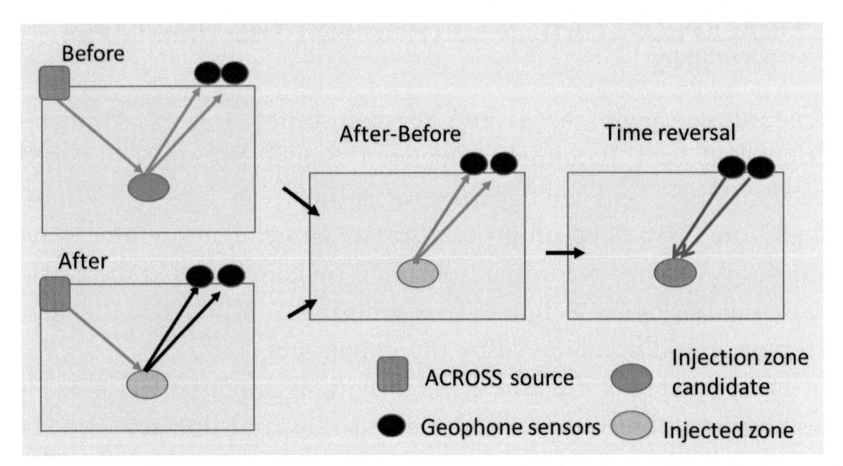

Figure 4.2 Principle of reverse time imaging. Subtraction of waveforms before and after the injection of vapor or supercritical CO_2 reveals the seismic waves radiated from the target zone to the geophones. The time reversal of residual waveforms is focused on the target zone.

We review the definition of Green's function, which is given in Section 3.1. We assume the single force $F_i(t, r_1)$ applied to the ground at the source r_1 in direction i, and the displacement $U_j(t, r_2)$ observed at the receiving point r_2 in direction j, where each i and j is either of x, y, or z. The transfer function between two locations $H_{i,j}(t, r_1, r_2)$ is given in Fig. 3.1 and Eq. (4.6),

$$U_j(t, r_2) = H_{i,j}(t, r_1, r_2) * F_i(t, r_1) \tag{4.6}$$

F_i and U_j are vectors and $H_{i,j}$ is a second-rank tensor. The asterisk denotes a convolution operator.

According to reciprocity theorem on Green's function, the force $F_j(t, r_2)$ at the receiving point r_2 in direction j gives the displacement $U_i(t, r_1)$ at the source r_1 in direction i.

$$H_{i,j}(t, r_1, r_2) = H_{j,i}(t, r_2, r_1) \tag{4.7}$$

Therefore we can use three-component observed waveform as a vector force input at each station for backpropagation. A similar approach has been done for the scalar wave field (Lumley, 2010). In addition to the backprojection of residual waveforms, cross-correlation of forward wave field and backpropagated wave field could enhance the imaging. This is examined in the Ketzin CO_2 storage case given in Section 4.3.

4.3 KETZIN CO_2 STORAGE CASE

4.3.1 Ketzin CO_2 Storage Experiment

Ketzin is located near Berlin, Germany. The injection was carried out into the Stuttgart Formation at ~630 m depth. The Stuttgart Formation is comprised of sands, sandstone, and mudstone. The shallower K2 reflector between the Weser Formation and Arnstadt Formation is identified by seismic reflection records and it dips toward the south (Juhlin et al., 2007). A former shallower gas-storage reservoir is between 200 and 450 m above the K2 reflector. During 2008 and August 29, 2013, the GFZ German Research Centre for Geosciences injected 67,271 t carbon dioxide into a saltwater-bearing sandstone at a depth of 630–650 m (Liebscher et al., 2012). Seismic time-lapse studies were carried out (Fig. 4.3) (Ivanova et al., 2012; Ivandic et al., 2012).

4.3.2 Model and Simulation

In order to do continuous monitoring of the behavior of injected CO_2 storage, the 4D seismic method is more expensive and cannot be carried out frequently. As the alternative method, Kasahara et al. (2013a) proposed to use the possibility of ACROSS as the seismic source and made the simulation of the Ketzin CO_2 storage study (Kasahara et al., 2013a).

Figure 4.3 Ketzin CO_2 injection site.

They modeled an N-S line above the Ketzin CO_2 storage site with the V_P, V_S, and density structure obtained by the Ketzin project. The CO_2 storage zone is located at 650 m depth and is 200 m wide and 10 m thick.

The simulation is similar to the previous one (Kasahara et al., 2010b, 2011a, 2012), but they modified the method to include the cross-correlation of the waveform from the source, similar to the reverse-time migration method used by Watanabe and Asakawa (2004). However, they used multisources for the imaging in cross-well shooting.

Calculated source gather waveforms using the finite difference method (FDM) program written by Larsen and Schultz (1995) are shown in Fig. 4.4 for no CO_2 storage and CO_2 storage. Very large amplitude surface waves are seen. The reflection from the K2 layer is seen at about 0.5 s two-way travel time (TWT). The reflection from the storage zone can be identified at 0.7 s TWT.

Residual waveforms between no CO_2 injection and CO_2 injection are shown in Fig. 4.5 and there are no surface waves that are common between the two sets of original waveforms. Both components show very distinct arrivals from the CO_2 storage zone.

In order to obtain a nice image, the cross-correlation of forward wave field and backpropagated wave fields was used. The time-reversed images shown in Fig. 4.6 were calculated by using the Z, X, and both components of residual waveforms (shown in Fig. 4.5) at every 40 m receiver spacing. However, the images are somewhat diffusive. If only the P portion of the Z-component is used, the image is sharper than the others. The storage zone is located between 0.7 and 0.9 km distance from the source and at 650 m depth. There is some signal at depths shallower than 400 m (Fig. 4.6: top), and it is generated by conversion of P and S waves. To eliminate this aliased image, the scalar field was used for the imaging. This gives a very sharp image and it is in the correct location in the model (Fig. 4.7). The final result of 2D imaging was compared to the assumed reservoir model (Fig 4.8). The result shows that the imaged and modeled reservoirs are placed at the exactly same location.

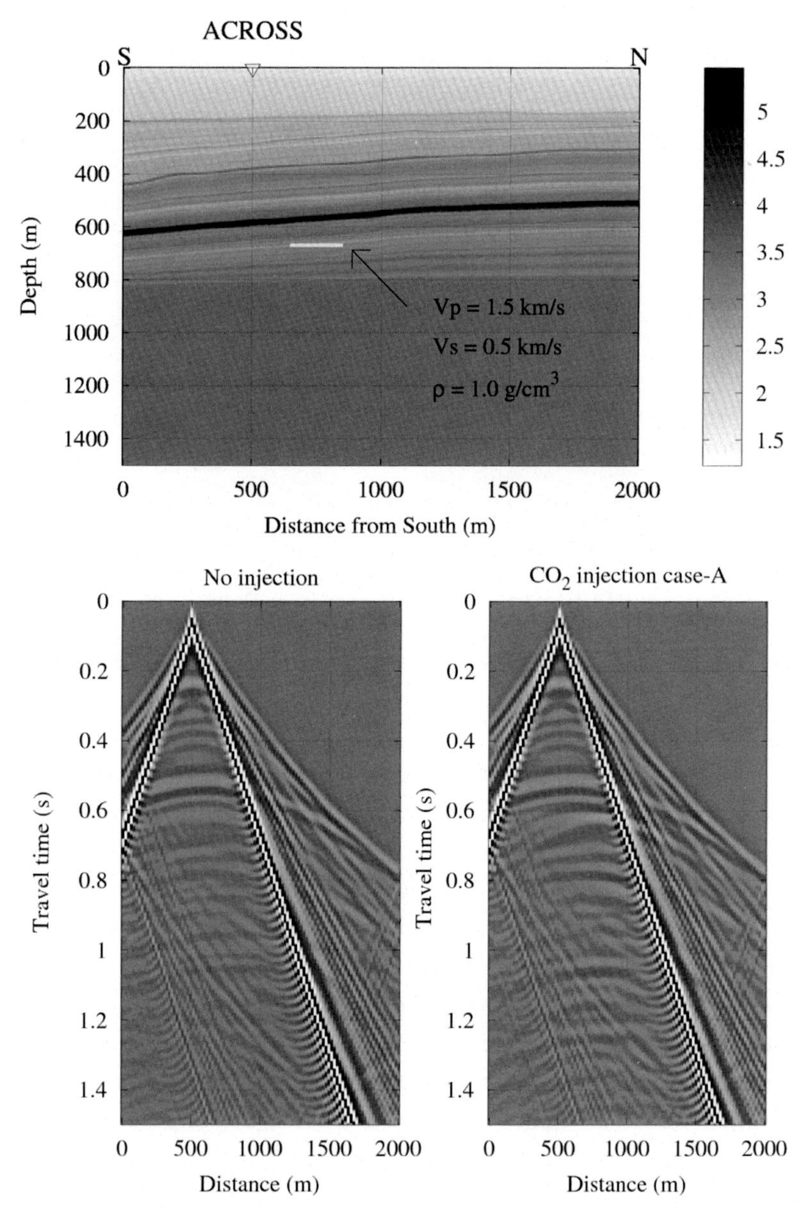

Figure 4.4 Velocity structure for the simulation (top) and waveforms from the ACROSS source for no injection (Z components) and injection (Z components). A 200-m-wide and 10-m-thick CO_2 storage is assumed.

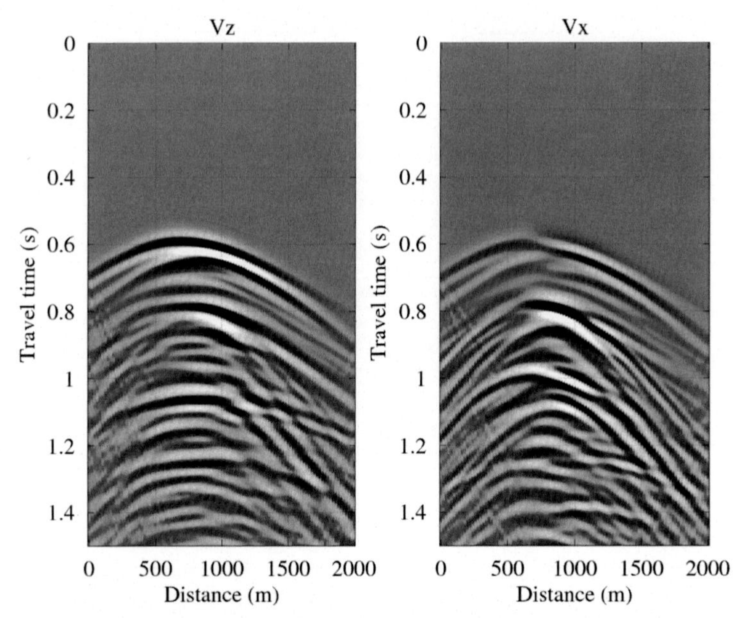

Figure 4.5 Residual waveforms between no CO_2 injection and CO_2 injection for Z-component and X-component. Injected area is 200 m wide and 10 m thick at 650 m depth.

However, the actual Ketzin test site has lots of constraints for the possible geophone locations. In order to examine the aperture effect, a 3D simulation assuming limited number of geophones in the array (Fig. 4.9A) was studied. The result of the 3D simulation is shown in Fig. 4.9B. The image is not clear because the receiver aperture is too small.

In order to enhance the image resolution, possible geophone locations in the Ketzin site were examined and it was found that possible locations for geophones were along an NW-SE road. If an additional line with larger aperture is added, the large aperture can greatly improve the image in depth and space (Fig. 4.10). This suggests that a large aperture is very important for imaging.

4.3.3 Conclusion in Ketzin Simulation

Assuming the geological and geophysical constraints of the Ketzin CO_2 storage experimental site, a time-lapse simulation was carried out. The simulation for the 2D case of an N-S profile gives a very

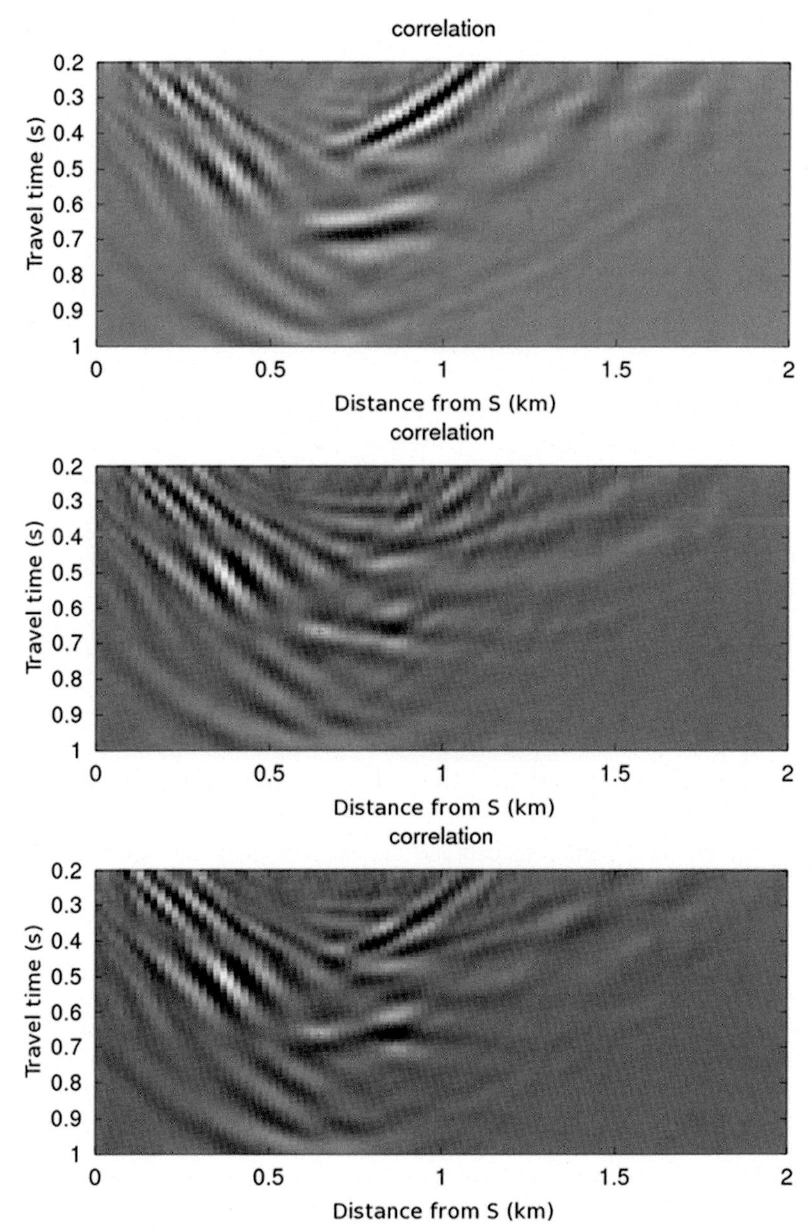

Figure 4.6 Imaging using residual waveforms of Z-comp. (top), X-comp. (middle), and Z- & X-comp. (bottom) by cross-correlation of source field. S: south (see Fig. 4.4).

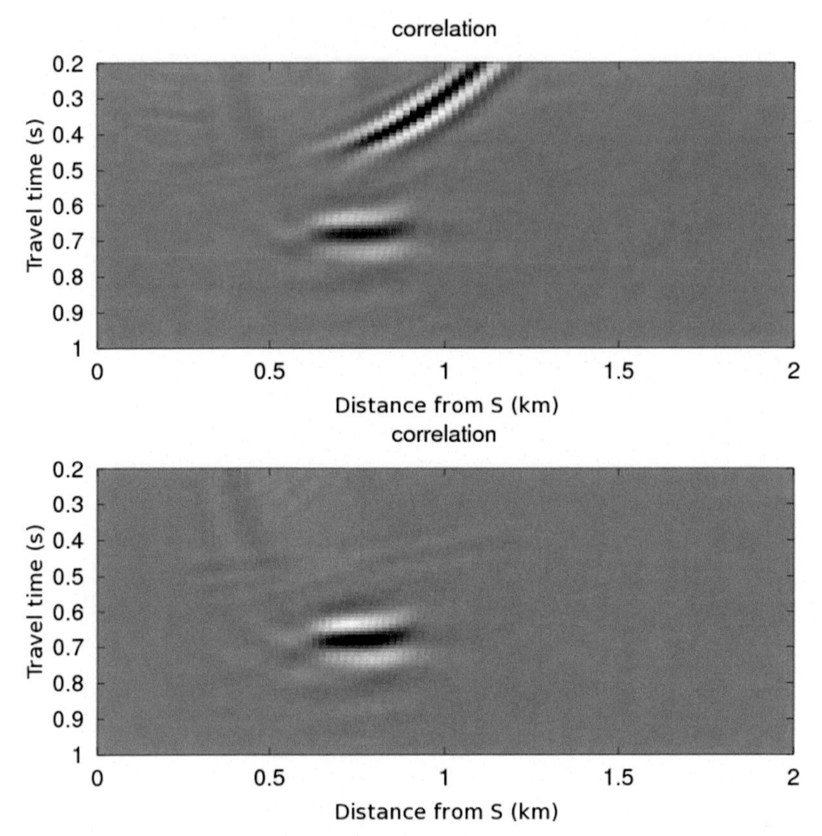

Figure 4.7 Use of P portion (upper) and calculation using scalar field (lower). The cross-correlation in scalar field gives a sharp image. S: south (see Fig. 4.4).

sharp image for the storage zone at 650 m depth with 40 m seismometer spacing. However, the simulation using realistic receiver geometry gives a less clear image for the CO_2 storage zone because receivers cover the limited area just above the storage zone. Additional coverage of receivers gives a better image resolution in depth and space.

4.4 OIL SANDS IN CANADA

The Japan Canada Oil Sands Limited (JACOS) Hangingstone operational area in Alberta, Canada, has used the steam-assisted gravity drainage (SAGD) technology to produce oil from oil-sand

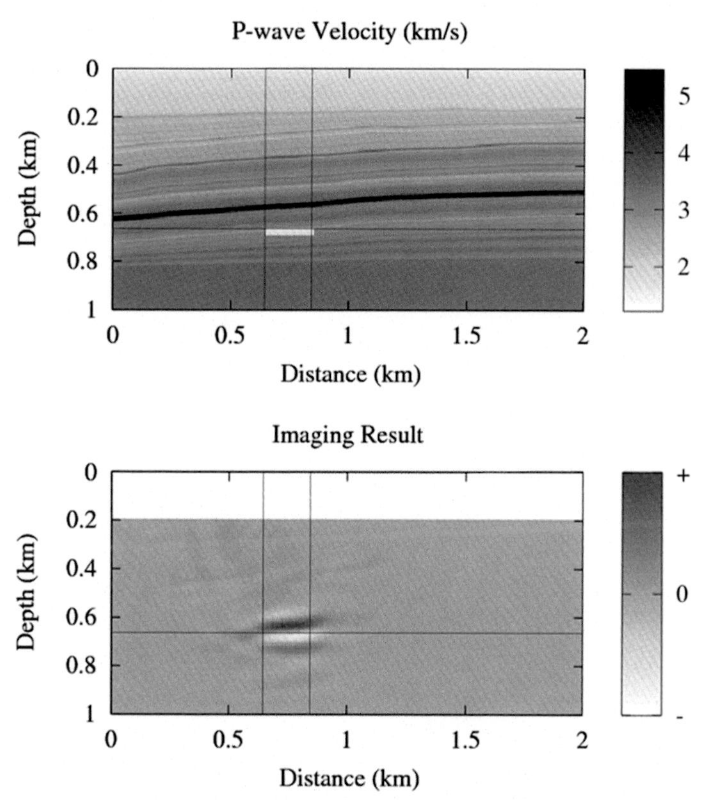

Figure 4.8 Simulation for CCS in Ketzin, Germany (Kasahara et al., 2013a). The vertical axis is depth in meters, and the horizontal axis is distance in meters. (Top) Model: The assumed injected zone (*white rectangle*) at 200 m in width, 20 m in thickness. *Gray scale* indicates V_P in km/s shown at the right-hand side. (Bottom) Result: assumed reservoir at 650 m in depth was clearly imaged. The thin lines represent the position of the target zone.

layers. JACOS has produced bitumen from oil-sand reservoir in the lower Cretaceous McMurray formation since 1999. The depth of oil sands is approximately 300 m. Bitumen tends to move as temperatures rose because its viscosity decreases with temperature increase. Nakayama et al. (2008) and Kato et al. (2008) studied the time-lapse 4D seismic monitoring for this field.

Using oil-sand samples from the JACOS field, Kato et al. (2008) measured the V_P and V_S by increasing temperature. Fig. 4.11 shows their results of the V_P and V_S with temperature under the pore pressure of 700 psi (= 4.762 MPa) and the confining pressure of

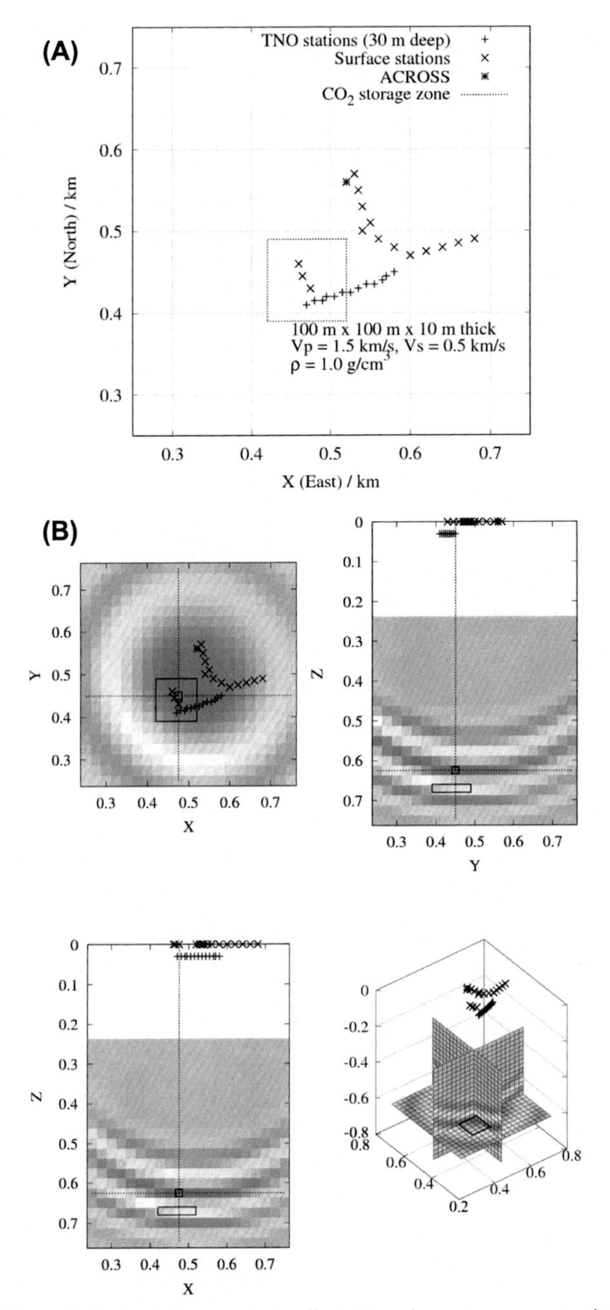

Figure 4.9 3D model to test the aperture effect. Geophones are assumed to consist of the existing TNO 30 m deep ones and some additional ones (A). Results (B): Top left: Depth slice at 620 m depth, top right: Depth-NS (Z-Y) section; Bottom left: Depth-EW (Z-X) section; bottom right: 3D image.

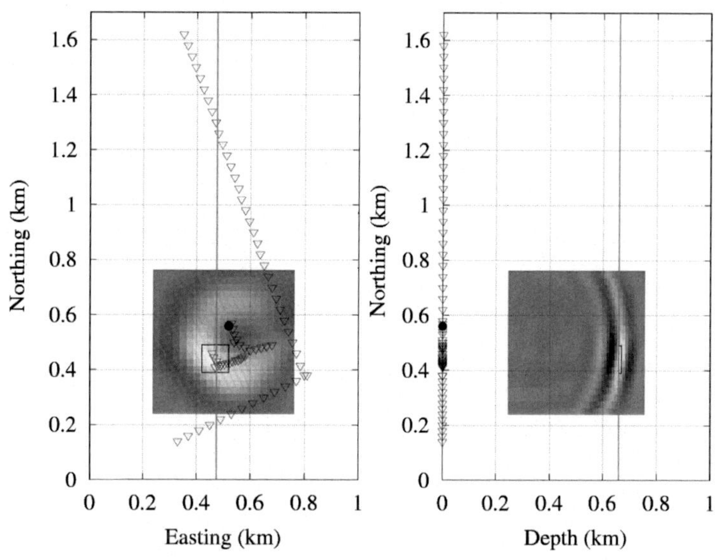

Figure 4.10 Addition of an NW-SE line and imaged zone (left) and improvement of depth resolution for the NS-depth (right).

900 psi (= 6.12 MPa). The maximum temperature of their experiment was 140°C and the V_P and V_S drastically decreased to 0.69 V_{P0} and 0.6 V_{S0} at 140°C relative to V_P at pore pressure of 300 psi (= 2.041 MPa) and 10°C, respectively. It should be carefully noted that the values of V_P and V_S by increase of temperature may or may not cause thermal cracks due to thermal expansion of mineral grains. In their experiment, the confining pressure of 900 psi seems a bit low to resist the thermal expansion. In addition to thermal expansion effects, dehydration of some hydrous minerals in oil-sand samples may occur at more than 100°C. The dehydrated water could affect the velocity measurement. Although effects of thermal expansion and dehydration may not be excluded in their experiment, the general tendency of velocity decreases with temperature increase can be acceptable. Because the temperature at oil production is approximately 260°C, the Kato experiment suggests the great decrease of V_P and V_S by the injection of high temperature vapor to the oil-sand layers if thermal expansion and dehydration effects did not occur in their experiment. Kato et al. (2008) compared results to the predicted velocity based on the rock mechanics and showed the

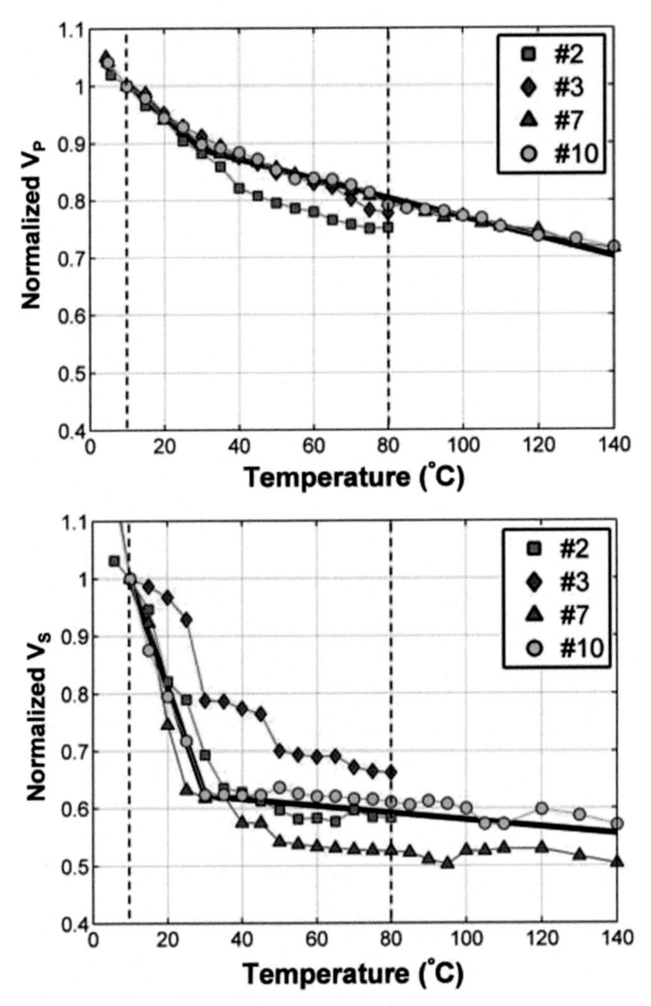

Figure 4.11 Normalized V_P and V_S of the oil sands with temperature at pore pressure of 700 psi (= 4.762 MPa) and confining pressure of 900 psi (= 6.12 MPa). *(After Kato, A., Onozuka, S., Nakayama, T., 2008. Elastic property changes in a bitumen reservoir during steam injection. The Leading Edge 27 (9), 1124–1131.)*

estimation using Gassmann's equation (Gassmann, 1961) at the temperature range of 0°C and 40°C was too low (Fig. 4.12). The temperature effects on rocks should be examined case by case.

Based on the experimental studied by Kato et al. (2008), 4D time-lapse study in the field experiment in Hangingstone operational area in Alberta, Canada, was conducted in 2002 and 2006 (Nakayama

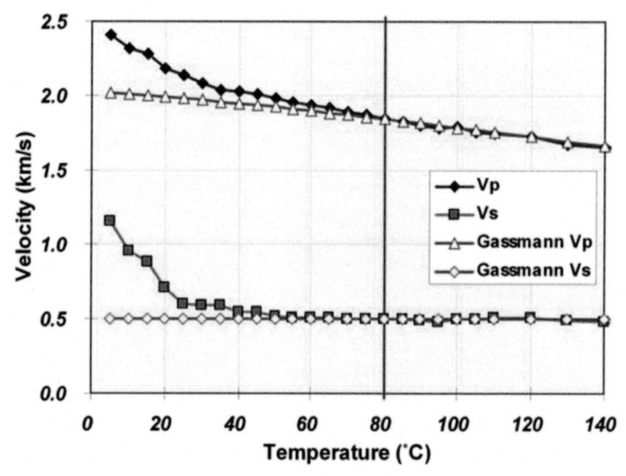

Figure 4.12 Comparison of measured and V_P and V_S and expected ones from Gassmann's equation of oil sands. *(After Kato, A., Onozuka, S., Nakayama, T., 2008. Elastic property changes in a bitumen reservoir during steam injection. The Leading Edge 27 (9), 1124–1131.)*

et al., 2008). In SAGD method, the injector of vapor is the upper-horizontal well and the producer of bitumen oil is the lower-horizontal well. The distance between the injector and the producer is 5 m. The 3D seismic data in 2002 and 2006 were compared (Figs. 4.13 and 4.14). There is distinct change between two 3D surveys in 2002 and 2006. The top Devonian at the southern half of the NS reflection line showed distinct lower bending by the decrease of velocity (Fig. 4.13). The travel-time changes occurred at the whole area of the western wells of H-Q (Fig. 4.14) and it is consistent with the change of lower bending of the top Devonian reflector. The amplitude difference concentrated in a much narrower region in the field (Fig. 4.15). Among time differences, amplitude differences, interval velocity changes, and cross-correlation of waveforms, the correlation between amplitude difference and waveform correlation is close to linear relation.

In the time-lapse studies in Hangingstone operational area in Alberta, Canada, by Nakayama et al. (2008) and Kato et al. (2008), it was thought that the time lapse of SAGD might be a good place to demonstrate the ACROSS time-lapse method using a single seismic source and a seismic array to image the physical state of bitumen.

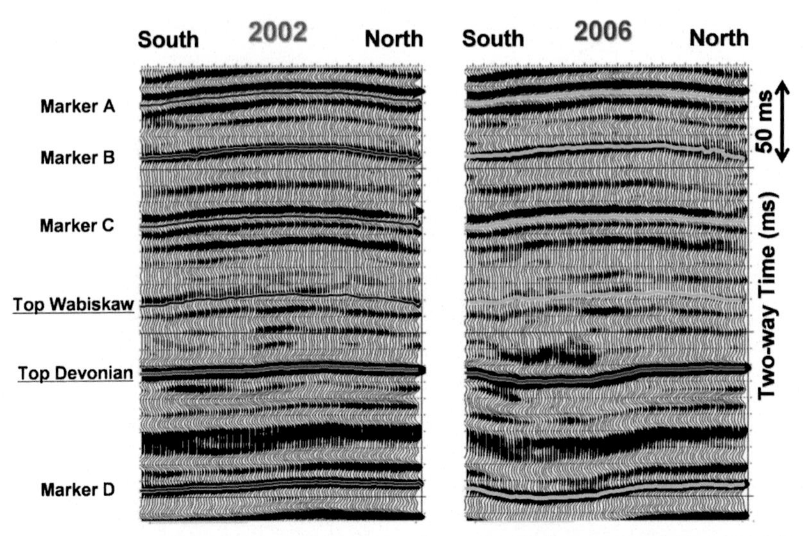

Figure 4.13 Interpreted seismic reflections along the NS line in 2002 and 2006. Top Devonian is regarded as the reservoir bottom (Base McMurray) and Top Wabiskaw is about 5 m shallower than the reservoir top (Top McMurray). *(After Nakayama, T., Takahashi, A., Kato, A., 2008. Monitoring an oil-sands reservoir in northwest Alberta using timelapse 3D seismic and 3D P-SV converted-wave data. The Leading Edge 27, 1158–1175.)*

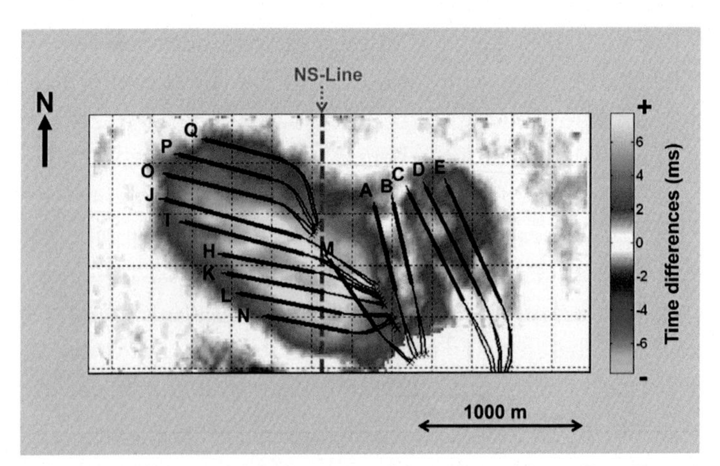

Figure 4.14 Time differences (ms) between the 2002 and 2006 Top Devonian horizons. The positive time areas show that the 2006 Top Devonian horizon is deeper than the 2002 Top Devonian horizon in the two-way time domain. *(After Nakayama, T., Takahashi, A., Kato, A., 2008. Monitoring an oil-sands reservoir in northwest Alberta using timelapse 3D seismic and 3D P-SV converted-wave data. The Leading Edge 27, 1158–1175.)*

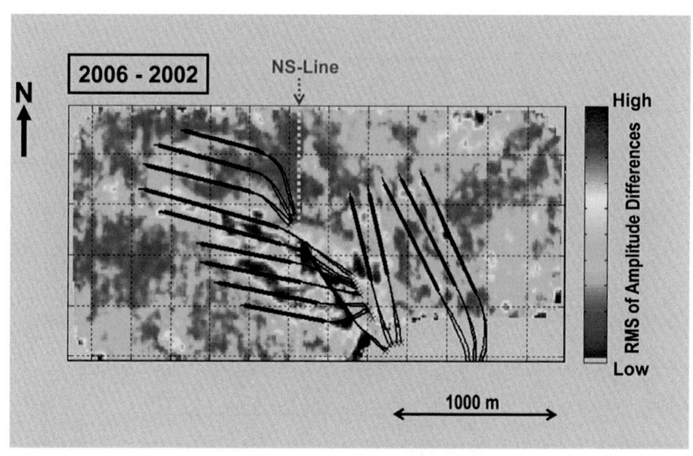

Figure 4.15 Results of JACOS SAGD time lapse. The rms values of trace amplitude differences within the reservoir on each trace location. Warm colors mean higher rms values of amplitude differences, and cold colors represent lower values. *(After Nakayama, T., Takahashi, A., Kato, A., 2008. Monitoring an oil-sands reservoir in northwest Alberta using timelapse 3D seismic and 3D P-SV converted-wave data. The Leading Edge 27, 1158–1175.)*

The following section gives the ACROSS time-lapse simulation study in the SAGD field in Hangingstone operational area in Alberta, Canada. Kasahara et al. (2013b) tested the possibility of time-lapse using single seismic source and geophone array. A similar very shallow reservoir case is also discussed in Section 4.6.

Considering results of Nakayama et al. (2008), the area of hot bitumen reservoir is modeled (Fig. 4.16). Five source locations (A–E) were examined for the imaging in the 3D model. The residual waveforms for source location A are given in Fig. 4.17. Because the bitumen reservoir is the left side of source A, the large residuals appear at the left side. Using only P portion of residual waveforms in Fig. 4.17, the backpropagated image is given in Fig 4.18. Among the images obtained by source A, B, and E, the vertical force at B (at the lower right in Fig. 4.18) gives the best image showing the model is almost recovered. Considering this simulation, the selection of the source location is important to obtain the good coverage, especially for shallower target case.

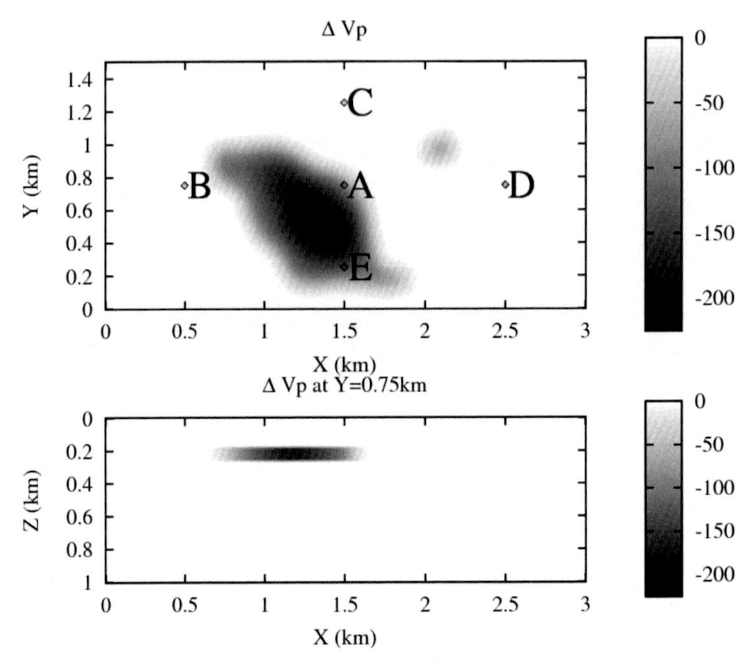

Figure 4.16 (Top) Top view of JACOS SAGD time-lapse model. (Bottom) Vertical profile of the model. A–E are assumed as the source positions.

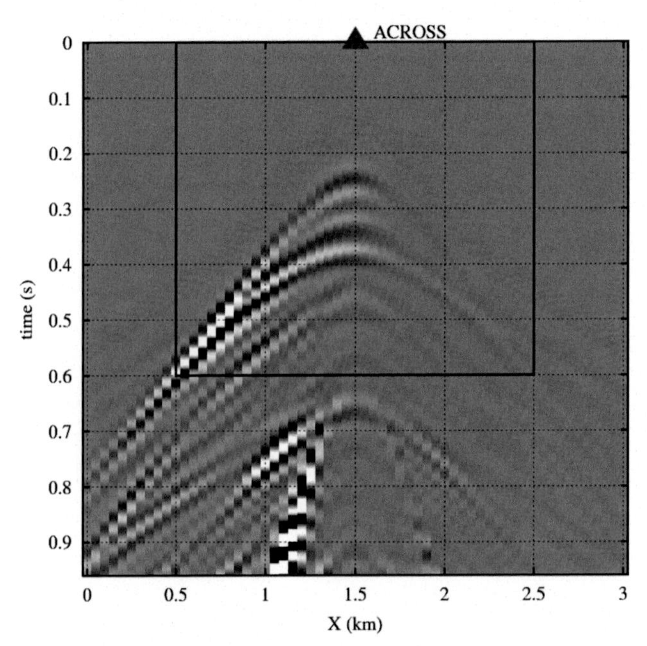

Figure 4.17 Residual waveforms along the EW line. Triangle is the location of source. Vertical axis is travel time. Waveforms in the rectangular part were used for the backpropagation imaging.

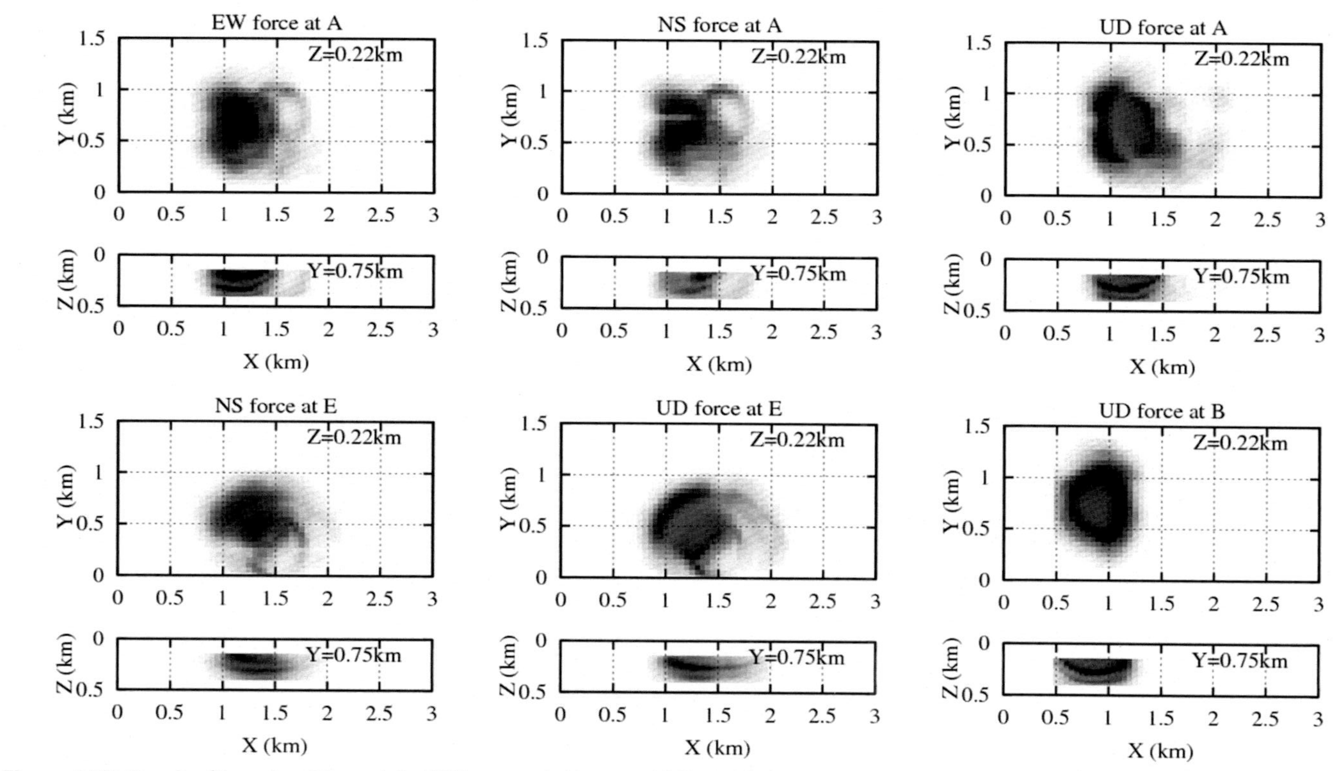

Figure 4.18 Result of imaging. Upper left: EW force at A; Upper middle: NS force at A; Upper right: vertical force at A; Lower left: NS force at E; Lower middle: vertical force at E; Lower right: vertical force at B (Kasahara et al., 2013b).

4.5 SIMULATION OF RESERVOIR AT 2 KM DEPTH

In this section, the effect of receiver interval for imaging is examined in 2D and 3D space. Conventional seismic reflection survey in 2D or 3D uses dense shot and/or receiver intervals at most 25 m. However, the denser intervals introduces higher costs to the seismic survey. In particular, the time-lapse requires semipermanent receiver installation, and if the sparse receiver interval gives satisfactory resolution, it is good news.

In Sections 4.5 and 4.6 the reservoir depths of deep and shallow cases were assumed. We synthesized 3D wave fields excited by vertical and horizontal forces by using FDM. The mesh size of finite difference calculation was 5 m and the source-time function was 20 Hz Ricker wavelet for both cases. The residual waveforms were calculated by synthetic waveforms before and after the changes in physical properties of the presumed reservoirs for use in imaging of temporal change.

4.5.1 Model and Simulation

In the 3D models with a reservoir at 2 km in depth as shown in Fig. 4.19, we assumed only one seismic source and a geophone array with 200 m spacing. The dimensions of the models were 3.5 km in X, 2.5 km in Y, and 4.5 km in Z.

By using this structural model in Fig. 4.19, we calculated V_{zz} waveforms before and after the injection of vapor or supercritical CO_2 into the heavy oil reservoir (Fig. 4.20), where V_{zz} represents the response to vertical force observed by the vertical geophone. We calculated the residual waveforms before and after the injection, which were used for the imaging (Fig. 4.20). Imaging was made by prestack Kirchhoff depth migration using the residual waveforms instead of backpropagation method. We used smoothed version of input velocity structure to calculate travel times for migration.

4.5.2 Results

The results of 3D simulation obtained by using the residual waveforms (Fig. 4.20) are shown in Fig. 4.21. It shows the retrieval image

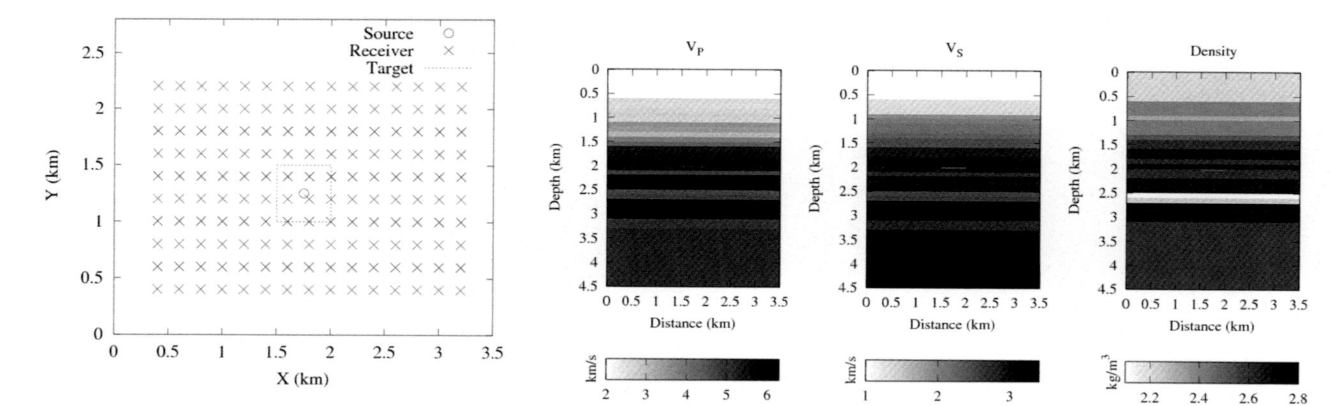

Figure 4.19 (Left) Reservoir of 500 m in X, 500 m in Y, and 20 m in thickness at (X, Y, Z) = (1.75, 1.25, 2 km). The receivers are at 200 m spacing grids. The seismic source is at (1.75, 1.25, 0 km). (Right) Structural model. The changes in physical properties in the reservoir are −5% in V_P, −20% in V_S, and −3% in density.

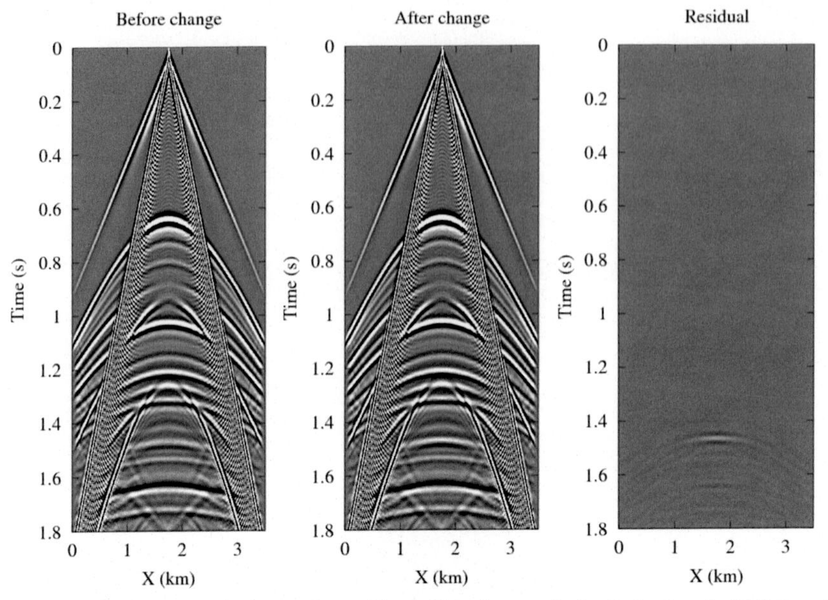

Figure 4.20 Source-gather waveforms V_{zz} (left) before and after injection (middle) and (right) residual waveforms (right). V_{zz} is the response to the vertical force observed by the vertical geophone.

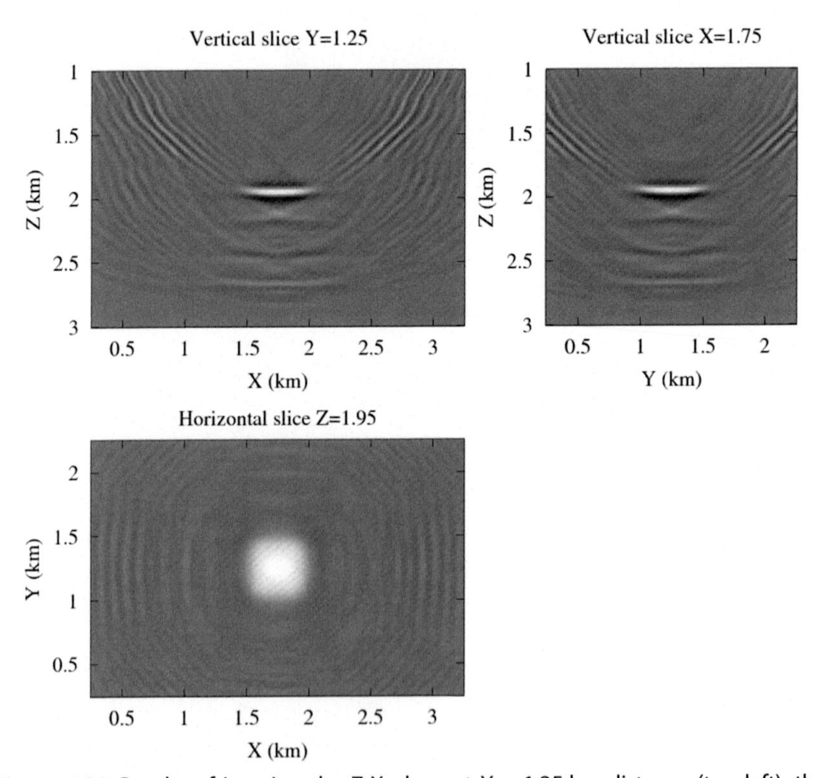

Figure 4.21 Results of imaging the Z-X plane at Y = 1.25 km distance (top left), the Z-Y plane at X = 1.75 km distance (top right), and the X-Y plane at Z = 1.95 km depth (bottom left) by using 150 geophones with 200-m spacing. The seismic source was only one at X = 1.75 km and Y = 1.25 km.

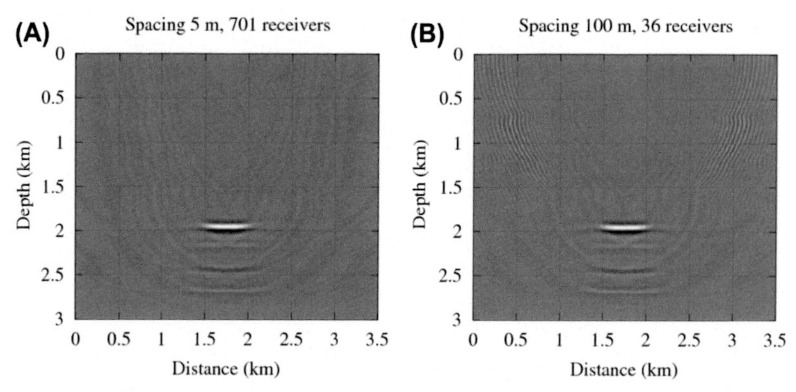

Figure 4.22 Comparison of receiver spacing effectiveness by using 3D forward waveform calculation and 2D imaging. (A) 5-m spacing; (B) 100-m spacing. The reservoir model in this simulation is 500 m long in X, 20 m thick in Y, and at 2 km depth. The results show no significant difference between images using geophone spacing of 5 and 100 m.

when using only one seismic source and 200-m-spaced 150 geophones were used. The result indicates that almost the same image of the model in Fig. 4.19 is retrieved even though the geophone spacing is 200 m.

Fig. 4.22 shows the effectiveness of the sparse geophone interval examined for 2D analysis by linear array. Although geophone intervals are 5 m and 100 m, we were unable to identify any significant differences between the two cases. If we can use 100 m spacing of geophones, the instrumental costs for the total installation costs of the monitoring system can be dramatically reduced.

4.6 SIMULATION OF VERY SHALLOW RESERVOIR

As we showed with the simulation of 300-m reservoir in Canada oil sands in Section 4.4, it is difficult to get a nice image with the reservoir model at shallow depth. In the real oil field, EORs at the much shallower heavy oil reservoirs have been in production. Considering the situation in the real world, we examined the case of a reservoir at 200 m depth. In addition, the injection wells intervals are so close, such as 100 m, and the size of each reservoir is extremely local.

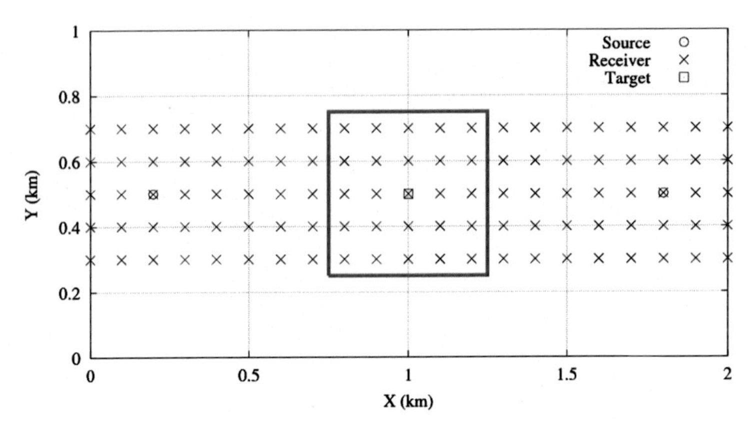

Figure 4.23 Shallow heavy oil reservoir model at 0.2 km in depth. The source location is at X = 0.2 km, and Y = 0.5 km. A small reservoir of 20 m in X, 20 m in Y, and 10 m in thickness is located at X = 1 km, Y = 0.5 km, and Z = 0.2 km. We calculated the image for assumed model by using 105 receivers and 25 receivers (represented by squares with thick gray edge). Physical properties used in simulations are given in Table 4.1.

4.6.1 Model and Simulation

We assumed a small reservoir of 20 m in X, 20 m in Y, and 10 m in thickness at 200 m depth (Fig. 4.23). In order to reduce the installation costs of equipment, as requested by oil production companies, we tested 100-m geophone grids and a seismic source to image the target reservoir. Imaging resolution of 25 m in the horizontal plane was also required. To judge whether the strict requirements are satisfied, we conducted 3D simulation. The dimensions of the model were 2 km in X, 1 km in Y, and 1 km in Z. Residual waveforms were calculated by subtraction of values before and after the injection of vapor. Imaging was made by prestack Kirchhoff depth migration using the residual waveforms instead of backpropagation method.

Table 4.1 The physical properties of the shallow heavy oil reservoir model

	Depth (m)	V_P (m/s)	V_S (m/s)	Density (kg/m^3)
Top layer	0–200	2500	1250	2.2
Bottom layer	200–1000	4500	2650	2.2
Reservoir (20 m × 20 m × 10 m)	190–200	2000	670	2.0

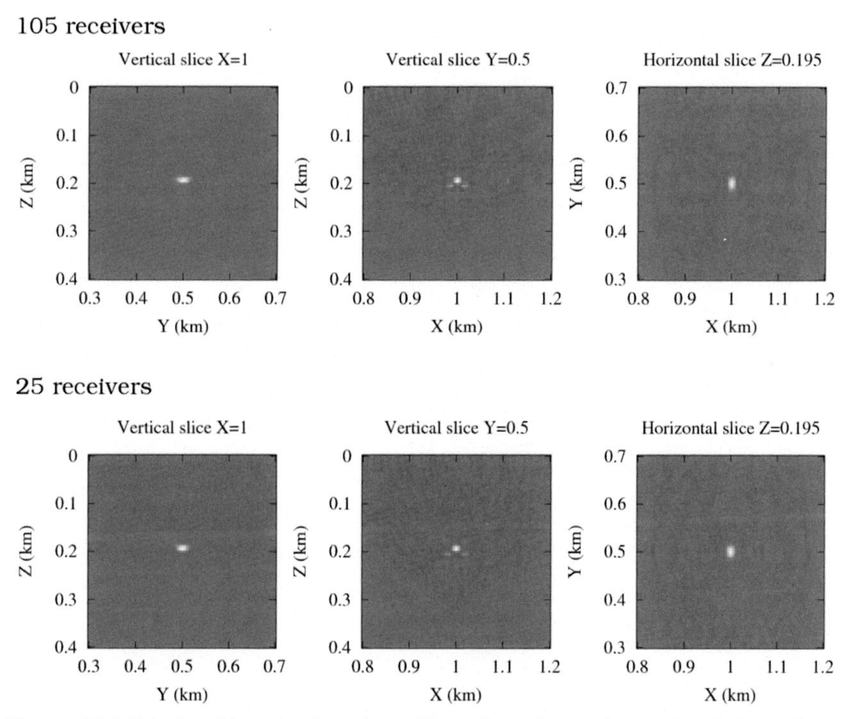

Figure 4.24 Results of imaging by using 105 receivers (upper images) and 25 receivers (lower images). Neither results show significant changes.

4.6.2 Results

The results of this simulation are shown in Fig. 4.24. Because the reservoir of this case is very shallow, the aperture angle might affect the imaging. Combination of migration results for PP and SS reflections provides high-resolution image. In our simulation, even if the source was 800 m from the target reservoir, the tiny target was effectively imaged. The image obtained by 105 and 25 receivers did not show significant changes. The reason is thought that the retrieved images strongly depend on receivers just above the target. Much denser geophones and additional seismic sources can improve the resolution of the results.

CHAPTER 5

Passive Seismic Approach

Contents

In this chapter, we will describe the passive approach. There are two seismic approaches for the time-lapse problem. One is passive method and the other is active method. In the shale gas production, the passive method using monitoring of microearthquakes is mainly used. In the 4D seismic methods, using air guns offshore and seismic vibrators onshore are good examples of active methods. The time-lapse study using the Accurately Controlled and Routinely Operated Signal System (ACROSS) seismic source is another example of active method.

During the active source operation, seismic waves generated by active source are thought to be noise for the passive measurements and disturb monitoring small quakes. If possible, the simultaneous use of passive and active methods for the time-lapse study is helpful for the progress of subsurface changes. This is enabled by use of the methodology of ACROSS signal processing.

5.1 SEPARATION OF PASSIVE (BACKGROUND) SIGNAL FROM ACTIVE (VIBRATOR) SIGNAL

In Chapter 3, we showed that the repetitive frequency sweep (chirp) of vibrator source generates a set of line spectra. When a T-long dataset in time domain recorded by geophone set is transformed to the frequency domain by discrete Fourier transform (DFT), the spectra of this dataset become a set of line spectra with $\Delta f = 1/T$ frequency spacing, where T is the length of time window. If the repetition period of the active source is T_m, the DFT of the signal

Time Lapse Approach to Monitoring Oil, Gas, and CO_2 Storage by Seismic Methods
ISBN 978-0-12-803588-7
http://dx.doi.org/10.1016/B978-0-12-803588-7.00005-4
Copyright © 2017 Elsevier Inc. All rights reserved.

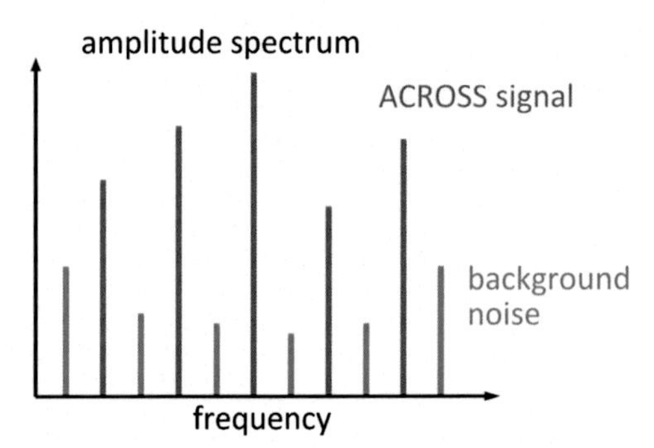

Figure 5.1 The spectrum distribution of signal and noise if the $2T_m$-long dataset are transformed to the frequency domain, where T_m is the length of time window of frequency sweep. Signal and noise spectrum distributes alternatively.

from active source is a set of line spectra with the spacing of $1/T_m$. Taking the data length $T = nT_m$ as an integer multiple of the T_m, the active signal components appear every n components in the discrete frequency series obtained by DFT. It should be noted that the separation of active signal from background noises in frequency domain is enabled only when the active signal is precisely repetitive.

The spectrum distribution of signal and noise is depicted in Fig. 5.1.

If background noises are fracking earthquakes, or quakes generated by brittle fractures, they can be used to determine the position of ongoing fractures.

5.2 MICROSEISMICS

In the shale gas production, the method of injection of liquid to generate brittle fractures in the shale layer is called fracking. The brittle fracturing might cause acoustic emission or microearthquakes. Microearthquake monitoring during the steam injection in steam-assisted gravity drainage was carried out by Maxwell et al. (2009).

The determination of hypocenters can express the place of ongoing fractures. The hypocenter determination of fracking noise is described in Eqs. (4.1)−(4.5). The origin time is included in these equations as unknown. Vesnaver and Menanno (2012) criticized the joint use of ground-surface and borehole data for the determination of origin time. It is necessary to apply the weathering correction to arrival times obtained by ground-surface geophones. Another point of view is the presence of anisotropy (Thomsen, 2013, 2015) because shale has large layering anisotropy up to 30%. There are two different types of anisotropy, that is, propagation direction anisotropy in P and S waves and polarization anisotropy causing S-wave splitting. The polarization anisotropy is caused by the layering of geological strata, the preferred orientation of constituent minerals, and the fabric of the fracture orientation. In addition to the anisotropy in shale, the differences of constituent minerals such as mica, illite, chlorite, kaolin, etc. also make the variations of V_P, V_S values and anisotropy (e.g., Sengupta et al., 2015). This affects the misallocation of hypocenters.

Moment tensor analysis and non-double-couple component analysis have been applied to the fracking quake analysis. During the first stage of fracking, volumetric expansion could occur at fractures. After the expansion of fractures, shear failures and double-couple acoustic emissions could occur. During the last stage, the fracture could close and implosion-type acoustic emission could follow. This process could be examined at the real field using non-double-couple moment-tensor analysis.

If it is necessary to do a passive seismic analysis during an active seismic source operation such as ACROSS, it is necessary to separate fracking quakes from the ACROSS seismic signals. This is similar to extraction of background noise from the ACROSS seismic source signal described in Appendix C.2. During the carbon capture and storage monitoring using ACROSS seismic source in Aquisotore, Canada, this simultaneous observation has been done by JOGMEC, Japan (personal communication).

5.3 SEISMIC INTERFEROMETRY

The principle of seismic interferometry was proposed by Claerbout (1968), Wapenaar (2004), Wapenaar and Fokkema (2006), and Wapenaar et al. (2010a,b).

$$\{G(x_B, x_A, t) + G(x_B, x_A, -t)\} * S_N(t) = \langle u(x_B, t) * u(x_A, -t) \rangle, \quad (5.1)$$

where $G(x_B, x_A, t)$ is Green's function between x_B and x_A, $S_N(t)$ is autocorrelation of noise, and $u(x_B, t) * u(x_A, -t)$ is cross-correlation of observed data at x_B and x_A.

In the theory, it is necessary to use random noise surrounding two stations. However, this condition was not achieved in the case of Al Wasse field, where ACROSS experiments were carried out. Although it was not perfect, Kasahara et al. (2014c) tested the seismic interferometry by limited circumstances. In this study, the noise during the ACROSS operation was extracted from the observed data based on the ACROSS processing.

Fig. 5.2 shows the ACROSS interferometry (Kasahara et al., 2014c). Four stations were used for demonstration as seen in Fig. 5.3. The results are shown in Figs. 5.4 and 5.5. The very strong stationary phase at 0.5 s was clearly seen. The main cause of ambient noise could be traffic noise from a nearby highway. Because the distance of two stations is 700 m, the apparent speed of this phase is 1.4 km/s. This phase seems to correspond to the surface wave.

It is noticed that the waveforms changed after January 19, 2013. Fig. 5.6 is the result of ACROSS as the source. Comparing the results of seismic interferometry using noise and time lapse using ACROSS as the active seismic source, similar waveform changes are identified. In this period, the outside temperature was below 0°C if the outside temperature at Al Wasse was similar to that at the Riyadh airport. From this result, the time lapse using seismic interferometry could be used.

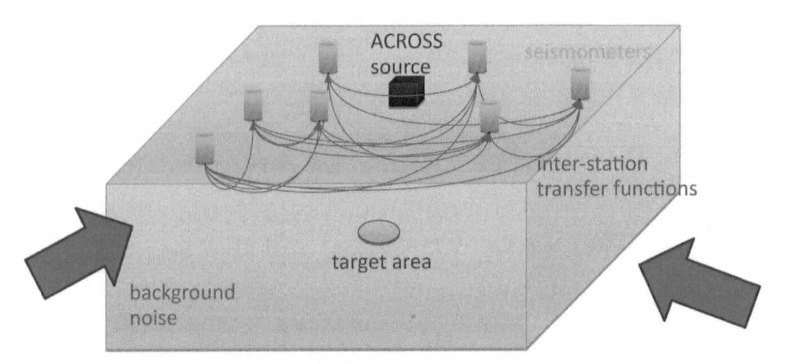

Figure 5.2 Cross-correlation of noise between two stations can extract seismic waves traveling between two stations.

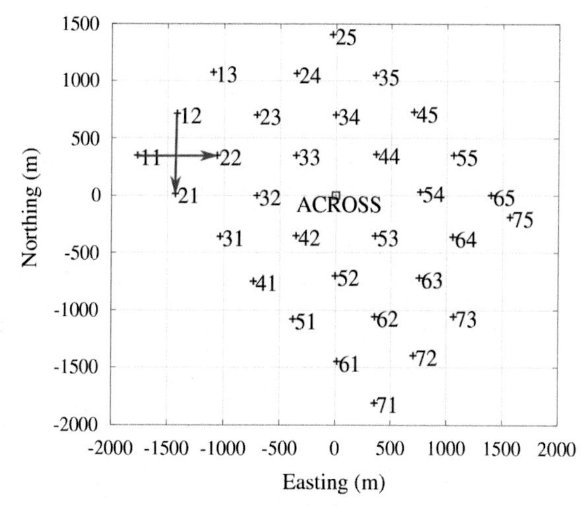

Figure 5.3 Seismic stations used for seismic interferometry. Noise components of four stations were used for interferometry. Both distances between #12 and #21 and between #11 and #22 are 700 m.

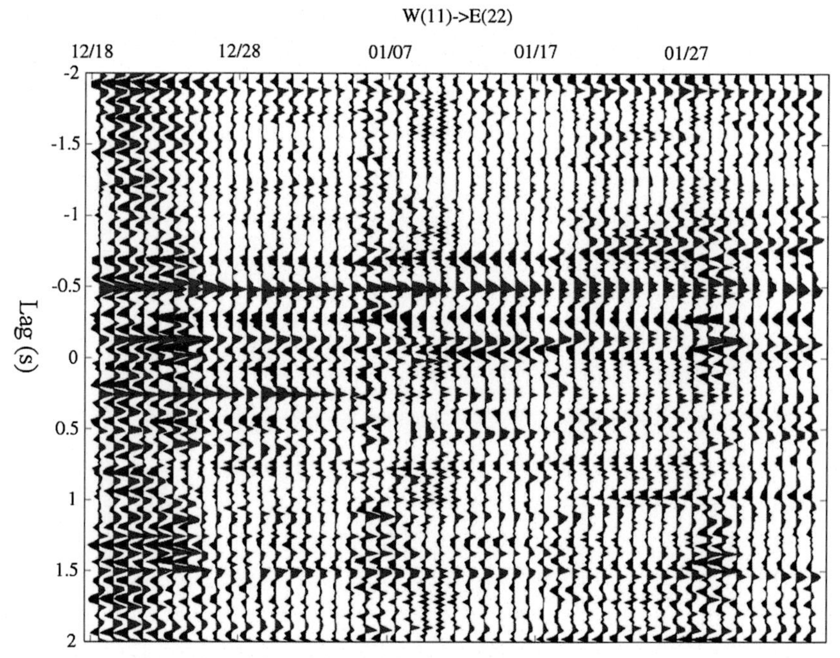

Figure 5.4 Seismic interferometry using cross-correlation of noise component of H_{VV} of two stations (#11 and #22). Forty-eight hours of data were used. The direction of #11 to #22 is positive time lag. Velocity of coherent phase is 1.4 km/s.

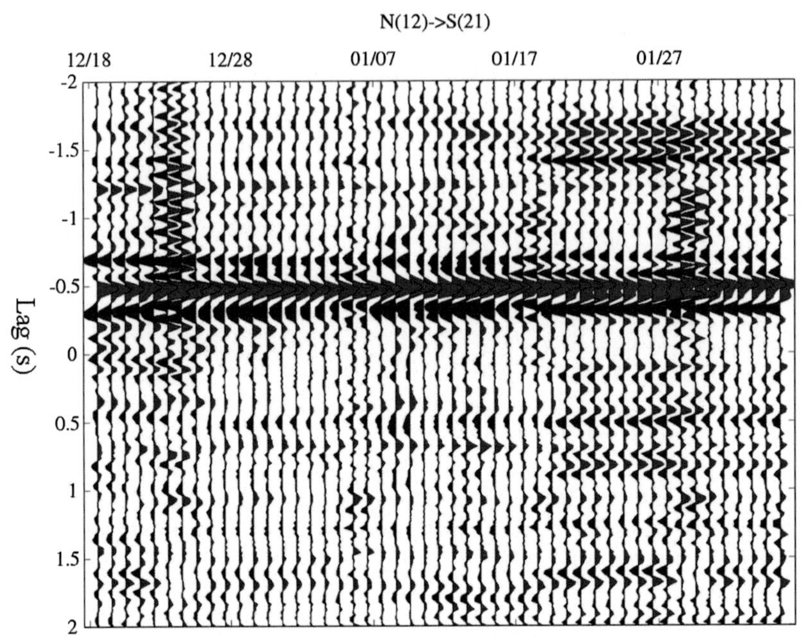

Figure 5.5 Seismic interferometry using cross-correlation of noise component of H_{vV} of two stations (#12 and #21). Forty-eight hours of data were used. The direction of #12 to #21 is positive time lag. Velocity of coherent phase is 1.4 km/s.

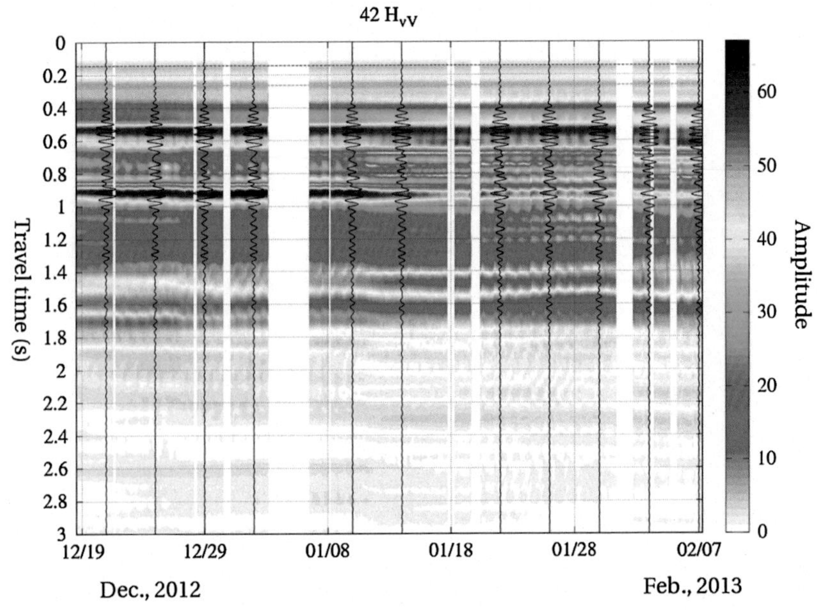

Figure 5.6 Results of active time lapse using ACROSS as the seismic source.

CHAPTER 6

Previous Time-Lapse Studies Other Than Accurately Controlled and Routinely Operated Signal System Method

Contents

There are so many carbon capture and storage (CCS) and enhanced oil recovery (EOR) projects in the world (https://sequestration.mit.edu/). Sleipner in Norway, In Salah in central Algeria, Weyburn in Canada, Nagaoka in Japan, Ketzin in Germany, and Otway in Australia are good examples for the time-lapse studies.

6.1 NAGAOKA CCS PILOT

The Japanese CO_2 pilot injection was carried out in Iwanohara of Nagaoka oil field owned by INPEX Co. between May 18, 2003, and January 7, 2005 (Fig. 6.1). The 10,400 t super critical CO_2 in total was injected into the Haizume formation which is an aquifer at the depth of 1100 m with 60 m thick. The critical point of CO_2 is 31°C and 7.4 MPa. The density of supercritical CO_2 is 656 kg/m^3. The IW-1 shown in Fig. 6.2 was a vertical injection well. The zone 2 of Haizume formation was selected as injection layer, which comprises fine to medium-fine sandstone and its permeability was 7 millidarcy

Time Lapse Approach to Monitoring Oil, Gas, and CO_2 Storage by Seismic Methods
ISBN 978-0-12-803588-7
http://dx.doi.org/10.1016/B978-0-12-803588-7.00006-6

Copyright © 2017 Elsevier Inc. All rights reserved.

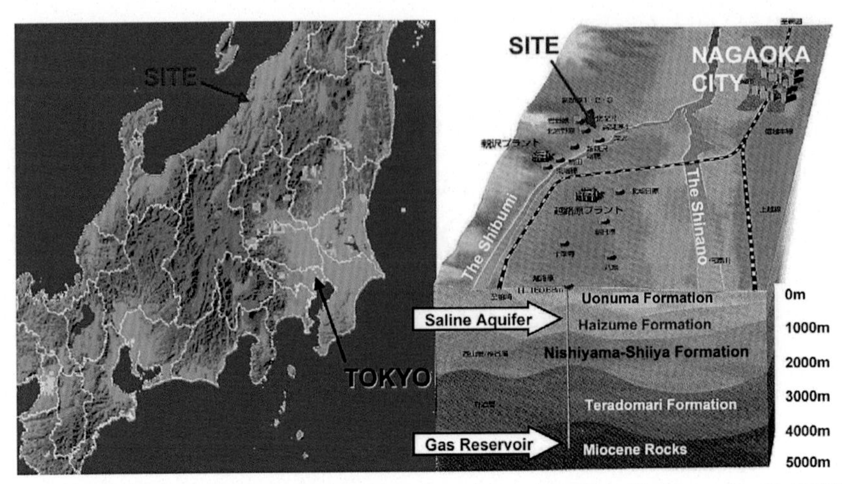

Figure 6.1 Nagaoka CO_2 injection site and geology. *After Xue, Z., Matsuoka, T., 2008. Lessons from the first Japanese pilot project o saline aquifer C2 storage. Journal of Geography 117 (4), 734–752.*

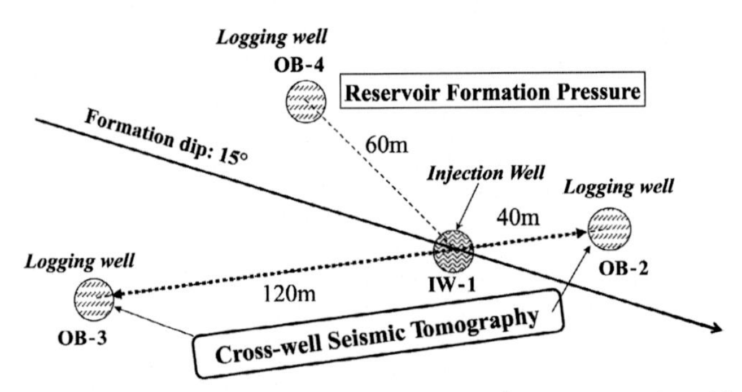

Figure 6.2 Map of wells configuration at the Nagaoka injection site. *After Xue, Z., Matsuoka, T., 2008. Lessons from the first Japanese pilot project o saline aquifer C2 storage. Journal of Geography 117 (4), 734–752.*

(mD, 1 darcy $= 0.986 \times 10^{-12}$ m^2). The injection was carried out from July 7, 2003, to January 1, 2005. The pressure and injection rate are shown in Fig. 6.3. V_P was measured by the sonic logging in the well 8 (Xue and Watanabe, 2008). The V_P before injection was 2.55 km/s and largely decreased to 1.84 km/s at the postinjection. The amount of velocity decrease was 28%. The thickness of velocity change was approximately 5 m distributing between 1113 m and 1118 m. The resistivity and neutron porosity measurement also showed similar change as ones of V_P.

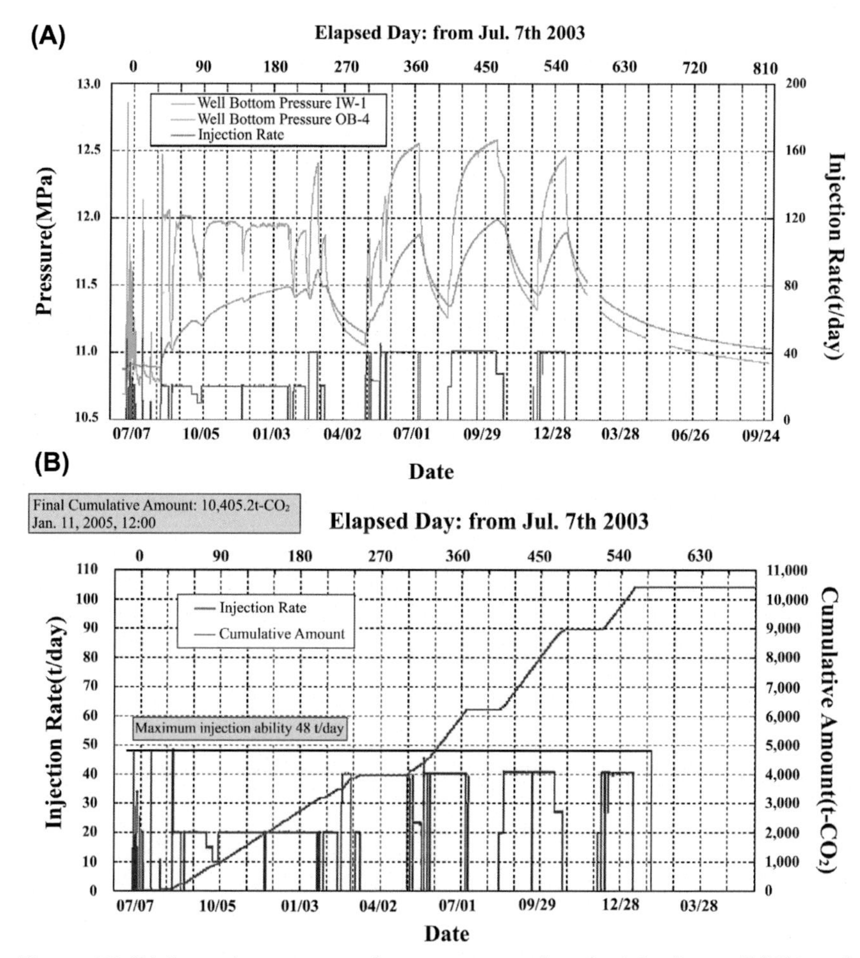

Figure 6.3 (A) Formation pressure changes measured at the injection well IW-1 and the observation well OB-4 during CO_2 injection and postinjection period. (B) CO_2 injection rate and total volume of injected CO_2 changes during the CO_2 injection period. *After Xue, Z., Matsuoka, T., 2008. Lessons from the first Japanese pilot project o saline aquifer C2 storage. Journal of Geography 117 (4), 734–752.*

The well logs and cross-well tomography, which were the first measurement in the world, were repeatedly done during the injection (Fig. 6.4). The V_P and resistivity could be decreased and increased by injection of CO_2, respectively (Xue and Ohsumi, 2004). The V_P measured by the sonic tool showed the change from 2.88 km/s before the injection to 1.84 km/s after the injection; the maximum change by injection was 28% (Xue and Matsuoka, 2008).

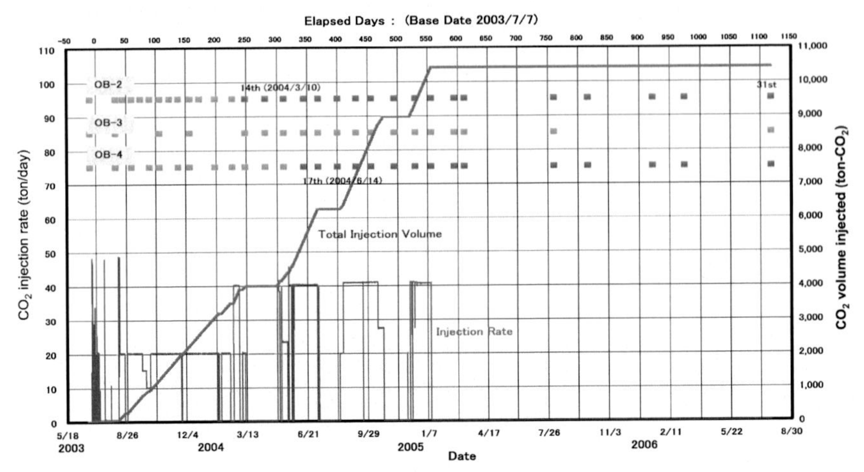

Figure 6.4 Time-lapse well loggings and cross-well seismic tomography done at the Nagaoka site. CO_2 breakthrough at the OB-2 and the OB-4 wells was detected at 14 and 17 loggings, respectively. Small squares (*Light gray*: prebreakthrough; *Dark gray*: postbreakthrough) and stars indicate runs for well logging and cross-well tomography, respectively. *After Xue, Z., Watanabe, J., 2008. Time lapse well logging to monitor the injected CO_2 at the Nagaoka pilot site. Journal of the Mining and Materials Processing Institute of Japan 124, 68–77.*

The porosity using Neutron log also shows 10% change associated with the CO_2 injection Figs. 6.5 and 6.6.

The cross-well seismic tomography was carried out between OB-2 and OB-3 during the injection stages of MS1–MS4. The seismic travel-time tomography based on ray theory showed a zone of velocity decrease, but the estimated change of V_P was 3.5% (Fig. 6.7) (Saito et al., 2008). The 3.5% of V_P change seems quite low compared to the sonic logging and core analysis, and they obtained 13.4% by constraint on the area of V_P change. Spetzler et al. (2008) obtained 18% V_P change by the waveform inversion technique.

The 3D seismic surveys were carried out twice at 2 × 2 km area before the injection in July 2003 and after the injection in July–August 2005, but the change of reflector by the CO_2 injection was not identified because the injection could be the limited source locations and receiver locations due to civilization area and the total amount of injection (Sakai, 2008; Xue and Matsuoka, 2008).

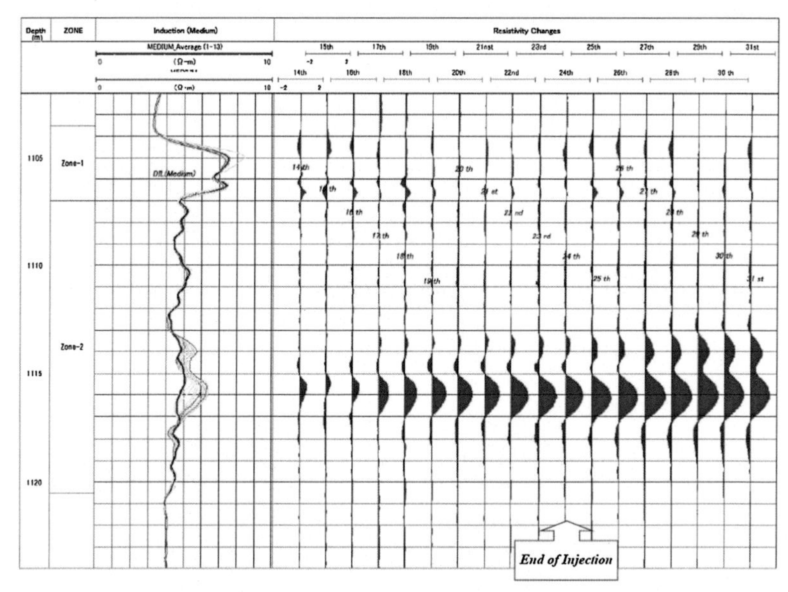

Figure 6.5 Results of induction log during the period of CO_2 injection and post-injection at the OB-2 well. The *blue line* in the left panel indicates the average of the logging runs 1–13. Logging charts in the right panel show differences from the average of 1013 data, indicating extension of the CO_2-bearing zone. *After Xue, Z., Watanabe, J., 2008. Time lapse well logging to monitor the injected CO2 at the Nagaoka pilot site. Journal of the Mining and Materials Processing Institute of Japan 124, 68–77.*

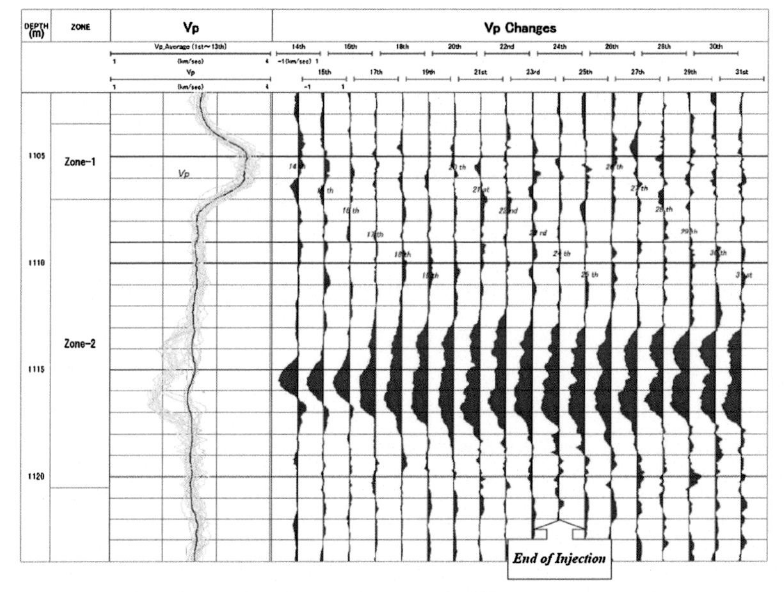

Figure 6.6 Results of sonic log during the period of CO_2 injection and postinjection at the OB-2 well. The *blue line* indicates the averages of P-wave velocity of runs 1–13. *After Xue, Z., Watanabe, J., 2008. Time lapse well logging to monitor the injected CO_2 at the Nagaoka pilot site. Journal of the Mining and Materials Processing Institute of Japan 124, 68–77.*

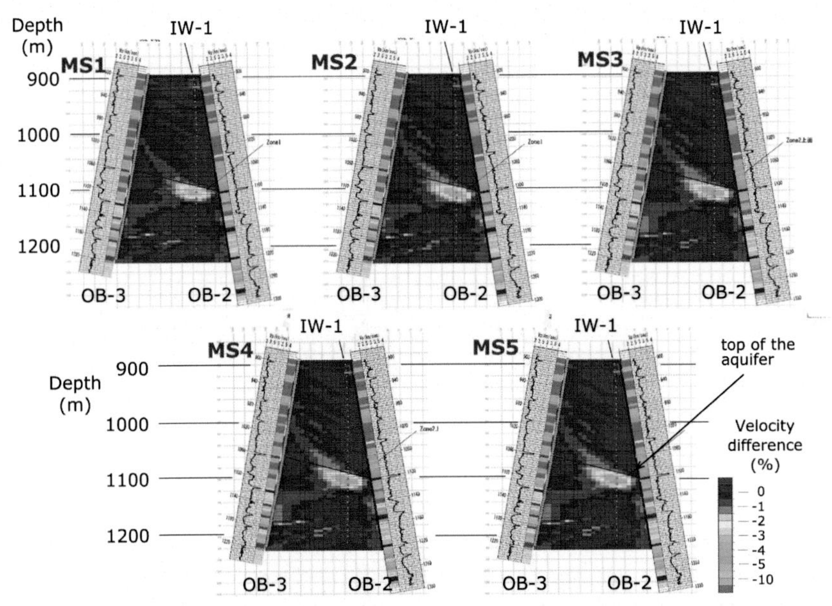

Figure 6.7 V_P difference tomograms for monitoring surveys of MS1 to MS4 during the CO_2 injection period. *After Saito, H., Nobuoka, D., Azuma, H., Xue, Z., 2008. Time lapse cross well seismic tomography monitoring CO_2 geological sequestration at Nagaoka pilot project site. Journal of MMIJ 124, 78–86.*

6.2 WEYBURN-MIDALE REGION (THE INTERNATIONAL ENERGY AGENCY GREENHOUSE GAS WEYBURN-MIDALE CO_2 MONITORING AND STORAGE PROJECT)

In Weyburn and Midale, Saskatchewan, Canada, the EOR by water flood was carried out, and the CO_2-EOR started in 2000 using CO_2 produced by the coal-gas plant in Coal Gasification and Coal Power from the Great Plains Synfuels Plant near Beulah, North Dakota, USA, at 330 km distance from Canada. CO_2 was transported by pipelines. The CO_2 miscible flood injection (e.g., Martin and Table, 1995) of 18 Mt in total and 2.8 Mt/y at Weyburn and 2.5 Mt in total and 0.46 Mt/y at Midale were conducted since 2000 (White, 2009; Whittaker and Wildgust, 2011). The injection zone in Weyburn was 30-m-thick fractured carbonate reservoir at the depth of 1400 m. The porosity and permeability are 15% and 10 mD in Weyburn and 16.3% and 7.5 mD in Midale, respectively (Koottungal, 2014).

EnCana runs the commercial EOR project and PTRC runs the research project looking at the potential to store CO_2. The 3D seismic surveys in 2001, 2002, 2004, and 2007 with the baseline survey in 1999 were carried out. The CO_2 injection area of Weyburn was clearly imaged by the time-lapse study using the 4D seismic surveys (Davis et al., 2003; White, 2009; Li, 2003). The differences of 3D seismic results showed clear change of P and S wave amplitude (Li, 2003; Davis et al., 2003; White, 2009). Davis et al. (2003) used multicomponent time lapse such as P-wave amplitude, travel-time difference, S-wave amplitude, and S-wave splitting. The results are shown in Figs. 6.8 and 6.9. Amplitude difference is larger than the travel-time difference (White, 2009).

The passive seismic measurement was also carried out, but little seismic activity was observed by the CO_2 injection. Between August 2003 and January 2004, 100 locatable earthquakes were detected with Mw (moment magnitude) = −3 to −1, which were concentrated near the injection well. The majority of the earthquakes showed low-frequency characteristics (White, 2009).

New CCS monitoring has been carried out at Aquistore near Weyburn, Saskatchewan, Canada. The time-lapse measurements

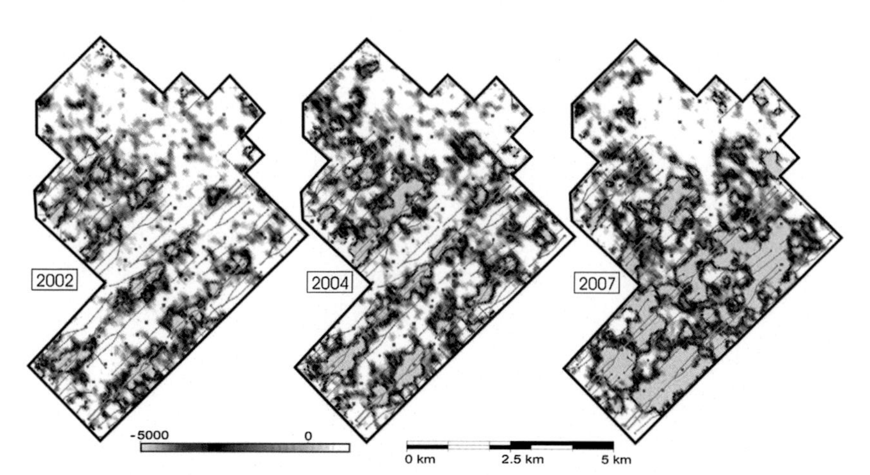

Figure 6.8 Results of 4D seismic study in Weyburn. Amplitude difference for the Midale Marly horizon from the baseline survey was mapped. Negative amplitude differences are CO_2 saturation. *After White, D., July 2009. Monitoring CO_2 storage during EOR at the Weyburn-Midale field. The Leading Edge, 838–842.*

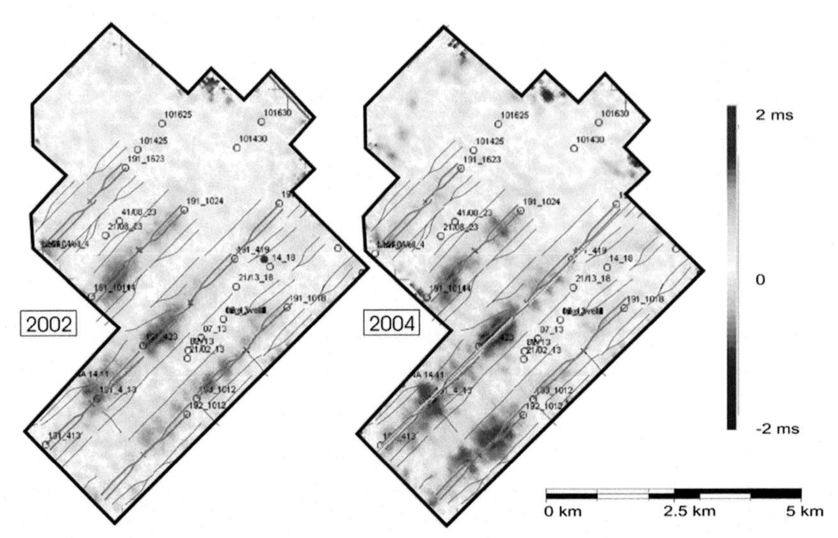

Figure 6.9 Results of 4D seismic study in Weyburn. Travel-time difference for the Midale Marly horizon from the baseline survey was mapped. *After White, D., July 2009. Monitoring CO_2 storage during EOR at the Weyburn-Midale field. The Leading Edge, 838–842.*

using 4D seismic and ACROSS time-lapse study have been carried out since 2014 (White, 2014). The CO_2 storage is deep saline aquifer storage (Winnipeg Sand/Deadwood Fm., 150 m thick) at 3.1 km depth. Reservoir cap rock is Icebox shale (15 m thick).

A number of downhole measurements have been carried out in Aquistore, that is, cross-well seismic and vertical seismic profile (VSP), cross-well and surface to downhole electrical monitoring, real-time pressure and temperature (P and T), passive seismic, fluid sampling, time-lapse logging, distributed acoustic/temperature sensors (DAS/DTS), and gravity. For active source and passive monitoring, 630 geophones over 6.25 km^2 at 20 m depth have been used.

6.3 IN SALAH

In central Algeria, In Salah, CO_2 separated from natural gas has been injected in the aquifer of Carboniferous formation since 2004. In contrast to the offshore CCS project in Sleipner in the North Sea and

Snøhvit in the Barents Sea, the CCS project in In Salah is onshore. The thickness of overburden Carboniferous and Cretaceous sequences are 950 m and 900 m, respectively (Ringrose et al., 2009). The depth of storage is 1900 m. The amount of CO_2 injection is 5–10% of natural gas. The permeability and porosity of In Salah storage zone is relatively very low (~10 mD) and 13%, respectively. For example, the permeability and porosity of Sleipner are ~3000 mD and 37%, respectively. Three long horizontal wells were drilled to improve injection rate. During the injection, considerable rise of reservoir pressure was observed. Gas has been produced from five wells with thickness of 20–25 m reservoir, and amine has been removed from gas and CO_2 was injected from three wells into Carboniferous sandstone at the Krechba field (Ringrose et al., 2009). Until 2013, 3.8 Mt CO_2 was stored in subsurface (Ringrose et al., 2013). The status of storage was measured by a number of measurements; time-lapse seismic, microseismic, wellhead sampling, downhole logging, core analysis, surface gas monitoring, groundwater aquifer monitoring, and InSAR. The upheaval of ground was measured by InSAR.

6.4 CO_2-CRC OTWAY PROJECT

Scientific CO_2 injection experiment was carried out in Otway in Australia in 2006–2007. There were two stages. During the first stage, 66,000 t of CO_2-rich gas taken from the gas reservoir of CO_2 + methane was injected into the depleted gas reservoir of the Naylor gas field. This field is located onshore in Victoria, ~300 km west of Melbourne. Gas was transported to the injection well at 2 km distance, and CO_2/CH_4 gas mixture with the ratio 80:20 was injected to CRC-1 at 2100-m-depth layer.

During the second stage, carbon dioxide up to 10,000 t was injected to the aquifer at 1400 m depth.

During the first stage, the following seismic surveys were carried out (Pevzner, 2012).

- Three dedicated 3D land seismic surveys: baseline and two monitor surveys acquired in 2008, 2009, and 2010.

- 4D VSP survey in CRC-1 borehole consisting of baseline (2008) and monitor (2010) acquired simultaneously with the corresponding surface seismic surveys.
- Zero-offset, offset, and walk-away VSP surveys in Naylor-1 and CRC-1 boreholes.
- A number of repeated acquisitions of a 2D test line.

However, changes of the elastic properties in the aquifers located above the reservoir in case of unlikely (but possible) escape of the gas from the primary containment and its upward migration could be very large.

It seems the higher repeatability was required. The main factors controlling the repeatability are significant levels of ambient noise and seasonal variations in the near-surface conditions (Pevzner et al., 2011; Kinkela et al., 2011). Pevzner et al. (2011) showed the repeatability was over 20% when combination of the mini-buggy in 2008 and weight drop in 2007 were used. Although the repeatability level of the 4D VSP data estimated for the Waarre C level is significantly higher (with median NRMS value of ~15%) than for the surface seismic data, this level of repeatability is still insufficient for robust detection of the time-lapse signal as its magnitude is similar to that of the time-lapse noise.

6.5 SLEIPNER

In Sleipner, the North Sea, CO_2 separated from natural gas has been injected to the Utsira sand formation, which is unconsolidated sandstone layer and saline aquifer located at 800–1000 m beneath the seafloor. The permeability and porosity of Sleipner is ~3000 mD and 37%, respectively. The upper part of Utsira layer is covered by shale layer with 40 km in EW and 200 km in NS. Since October, 1996, the CO_2 has been injected by the near-horizontal well at the depth of 1012 m below sea level some 200 m below the reservoir top (Arts et al., 2004; Chadwick et al., 2010). The CO_2 injection was 1 Mt/y and more than 11 Mt in total (Chadwick et al., 2010). The repeated 3D seismic surveys were carried out in 1994, 1999, 2001,

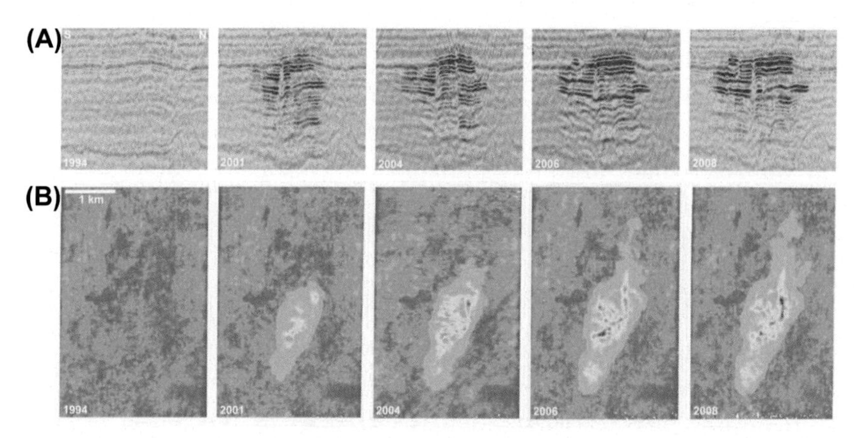

Figure 6.10 The time-lapse reflection records of the Sleipner CO_2 injection plume. (A) NS line and (B) top view of plume. *After Chadwick, A., Williams, G., Delepine, N., Clochard, V., Labat, K., Sturton, S., Buddensiek, M.L., Dillen, M., Nickel, M., Lima, A.L., Arts, R., Neele, F., Rossi, G., 2010. Monitoring Quantitative analysis of time-lapse monitoring data at the Sleipner CO_2 storage operation. The Leading Edge 29, 170–177.*

and 2002. The 4D seismic data were used for the CO_2 plume evolution (Arts et al., 2004, 2008; Chadwick et al., 2009, 2010). The plume was imaged as a number of bright horizontal reflections within reservoirs with time evolution (Fig. 6.10). In 2008, as seen in Fig. 6.10, the size of plume was ~3 km in NS and ~1 km in EW and the velocity pull down in the center of plume was seen (Chadwick et al., 2010). Ghaderi and Landro (2009) analyzed the amplitude and time-shift changes to estimate the velocity and thickness changes within CO_2 saturated layer. Clochard et al. (2010) used impedance change between 1994 and 2006. The P-wave impedance reflection section showed the push-down structure in the base of Utsira layer with the ~2000 m horizontal width, but the S-wave impedance change was weaker than P. The maximum push down of 35 ms of S-wave along the NS line was estimated to be distributed inside of the plume at the center of injection well. The dual-sensor streamer technology was used in Sleipner CO_2 injection monitoring to give the higher resolution by high-frequency contents of survey data (Furre and Eiken, 2014).

6.6 PERMANENT RESERVOIR MONITORING

Permanent reservoir monitoring (PRM) is one of the time-lapse studies. This method has been used extensively in the North Sea and off Azerbaijan and was initiated off Brazil. We described the PRM in Section 2.2.

6.7 OTHER AREAS

In Delhi Field, Louisiana, CO_2 flood was carried out (Davis et al., 2003; Richard, 2011; Bishop and Davis, 2014). They used the multicomponent time lapse, which was used in Weyburn, Canada (Davis et al., 2003).

The reservoir in Delhi Field consists of Upper Cretaceous (Tuscaloosa) and Lower Cretaceous (Paluxy) sandstones. In this area CO_2 flood EOR has been carried out in 1-km-depth layer. The late Cretaceous Tuscaloosa unconformity overlies the Paluxy whose thickness is from 9 to 23 m. The porosity and permeability of the injection layer is 25% and 1400 mD, respectively (Davis, 2015). P- and PS-wave amplitudes and P- and PS-wave time shifts were discussed (Bishop and Davis, 2014). Davis (2015) interpreted the migration of CO_2 using the RMS amplitude difference of PP and PS at top Paluxy layer. Time lapse of time shift occurred in the overburden layer by CO_2 injection. Although PS-wave amplitude change was weak and not related to the CO_2 injection well, P-wave amplitude showed strong changes related to CO_2 injection. PS-wave time shift was large showing compaction that occurred in the overburden above the CO_2 injection well (Bishop and Davis, 2014).

During the CO_2 injection, there are several areas doing monitoring, that is, Vaccum Field, Southeastern New Mexico, West Pearl Queen Field, and Postle Field (Melzer and Davis, 2010). In the Vaccum Field, multicomponent time lapse has been carried out. Fifty million standard ft^3 CO_2 injections from one well and one billion ft^3 injection from six wells were carried out and shear wave velocity anisotropy changes were observed (Melzer and Davis, 2010). The depth of injection zone was 6 m thick at 1364 m depth.

The CO_2 injection at West Pearl Queen Field, New Mexico, was carried out in the depleted sandstone layer with 6 m thick at 1364 m depth. Two high-resolution multicomponent seismic surveys done six months apart showed the shear wave anomaly (Melzer and Davis, 2010).

CHAPTER 7

Case Studies Based on Accurately Controlled and Routinely Operated Signal System Methodology

Contents

There have been many time-lapse studies in the past. In this chapter, we describe the field tests of Accurately Controlled and Routinely Operated Signal System (ACROSS) time-lapse experiments. Four field tests of the time lapse were carried out in central Japan, Awaji Island and

Time Lapse Approach to Monitoring Oil, Gas, and CO_2 Storage by Seismic Methods
ISBN 978-0-12-803588-7
http://dx.doi.org/10.1016/B978-0-12-803588-7.00007-8

Copyright © 2017
Elsevier Inc.
All rights reserved.

green tuff quarry in Japan, and in Saudi Arabia. For the field tests in the central Japan, Awaji Island, and in Saudi Arabia, the ACROSS seismic source was used. However, for the field test in the green tuff quarry in NE Japan, a modified conventional seismic source was used.

7.1 CASE STUDIES IN JAPAN

Using the ACROSS seismic source, people in Tono Geoscience Center of Japan Atomic Energy Agency (JAEA) and Nagoya University initiated the time-lapse experiment in central Japan and Awaji Island.

Using one ACROSS seismic source located in Toki city, Japan, Yoshida (Yoshida et al., 2010; Yoshida, 2011) carried out the time-lapse experiment to monitor the waveform changes during 1.5 years (February 20, 2004—August 13, 2005). He examined the waveform changes of the ACROSS signal at the Horai seismic station (N.HOUH) of the National Research Institute for Earth Science and Disaster Prevention transmitted from the JAEA ACROSS source with the vertical rotational axis (ACROSS-V) at Toki city. The frequency band was 15.47—20.57 Hz, which is narrower than one by the ACROSS source with the horizontal rotational axis (ACROSS-H) in Saudi Arabia. Because the offset distance between source and receiver was 57 km, it was necessary to stack the data during 10 days to obtain the reasonable signal-to-noise ratio (S/N) to compare the first arrivals for the long duration. He detected the seasonal variation on the later part of P-phase, but there was no significant temporal variation on the first break even though the station was placed at 200 m depth in the borehole. He suggested seasonal rain and/or temperature variation as the cause of temporal changes seen in the later parts. We think that one of possibilities is near-surface effects on the PP reflection at the surface even though the geophones were set at 200 m depth. The similar temporal changes were also seen in the later phases of borehole geophones at 200 m depth in NE Japan (given in the near-surface effects) described in Section 7.3.

Yamaoka's group (e.g., Misu et al., 2004) used an ACROSS-V source located in Awaji Island to try to find the time lapse associated with the earthquake generation. They also examined the ACROSS data during the water injection. When 250 tons water were injected to the borehole down to 650 m depth with 4.5 MPa between March 13 and May 25 in 2003, there were no significant temporal changes associated with the injection, but they found the correlation between arrival-time changes of S-wave and surface wave and precipitation (Misu et al., 2004). By analysis of water flow rate due to precipitation, Kasahara et al. (2012) showed the arrival-time delay seen in S- and surface waves showed the same pattern of flow rate by precipitation (see Chapter 8). In addition, they showed the concentration of residuals focused along the earthquake fault of the Great Hanshin earthquake (Mw = 6.9) in 1995 (Kasahara et al., 2012).

7.2 AIR INJECTION EXPERIMENT IN AWAJI ISLAND

7.2.1 Field Setting and Description of Field Test

Kasahara et al. (2013c) carried out a time-lapse experiment using two ACROSS seismic sources, 31 3C-geophones and air injection into the formation at 100 m depth in Awaji Island, where was the focal zone of the Great Hanshin earthquake in 1995. The velocity profile across Awaji Island was obtained by Sato et al. (1998). The present section describes the results of the time-lapse (4D) experiment as the above.

7.2.2 Injection Experiment and Data Processing

Kasahara and others (Kasahara et al., 2013c) used the ACROSS-H seismic source with the horizontal rotational axis in addition to another ACROSS-V seismic source with the vertical rotational axis. They placed 31 3C-seismometers at the ground surface in addition to 800-m-borehole 3C-seismometers. Locations of these are shown in Figs. 7.1. They operated both ACROSS sources. In this experiment, the ACROSS-V was operated by the 10–35 Hz sweep with 100 s window. In an hour, they repeated 32 sweeps and a 400 s transitional period. The rotation of the ACROSS-H was alternatively switched between clockwise and counterclockwise every 1 h. Synthetic seismic records for vertical and horizontal vibrations were calculated

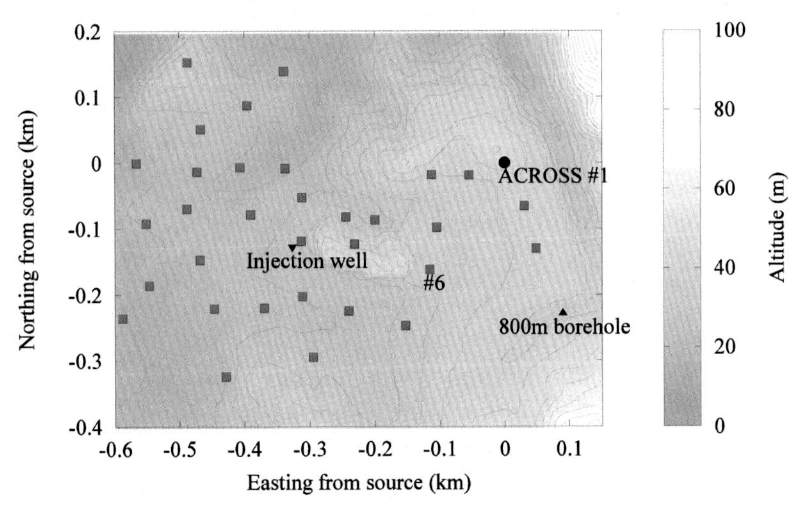

Figure 7.1 Location map of Accurately Controlled and Routinely Operated Signal System (ACROSS) #1 (*solid circle*), air injection well (*solid down-pointing triangle*), receivers (*gray squares*), and the receiver #6.

as described in Chapter 3. The transfer functions between the ACROSS source and each geophone on the surface and the 800-m borehole were obtained.

The 80 t air in total with 2.1 MPa pressure were injected into the quaternary sedimentary formation at 100 m depth between February 26 and March 3, 2011 (Fig. 7.2). At two days before the main injection, 1 t injection of air was injected as test. Although there are some differences in physical and chemical behaviors between CO_2 and air injections, the first objective of this field pilot test is to determine the migration of air—water in the ground to prove the effectiveness of their time-lapse method.

7.2.3 Observed Results

The vertical and horizontal vibration records were shown in Fig. 7.3, respectively. The observed seismic records corresponding to clockwise and counterclockwise rotations of the ACROSS seismic source were added or subtracted. The addition and subtraction give seismic records for the vertical force (UD) and the horizontal force (EW), respectively. Vertical and EW components were used for vertical and

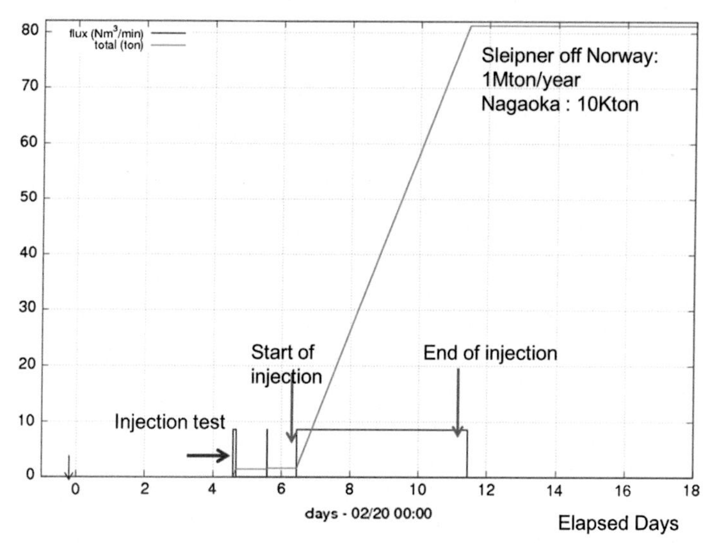

Figure 7.2 Time sequence of air injection. Accumulated air weights and air flow/min (vertical axis) with elapsed days (horizontal axis).

horizontal force responses, respectively. In this figure, P-wave with c.a. 2.5 km/s is identified in the UD records of vertical vibration, and S-wave with 400 m/s is identified in the EW records of horizontal vibration.

The residual waveforms in reference to the seismic record at 0:00 on February 24 were calculated and the results are shown with original waveforms (Figs. 7.4 and 7.5). The 800-m borehole data are shown in Fig. 7.5. The slant distance between the source and seismometers at 800 m is ~0.81 km and V_P is estimated as approximately 4 km/s, which is the one of weathered granite. It is surprising that the residual waveforms show large arrivals after P arrivals immediately after the air injection. This observed rapid response can be explained by the presence of macroscale fractures at E—W direction rather than extremely large permeability at this depth. The borehole geophone showed some residual arrivals before the main injection, but it seems the response to 1 t test injection. All seismic records show extremely large changes on residual waveforms after the start of air injection. Some stations show immediately after the air injection, but some stations such as stations #6 and #7 show gradual changes on waveforms (Figs. 7.4 and 7.6). Because the waveform

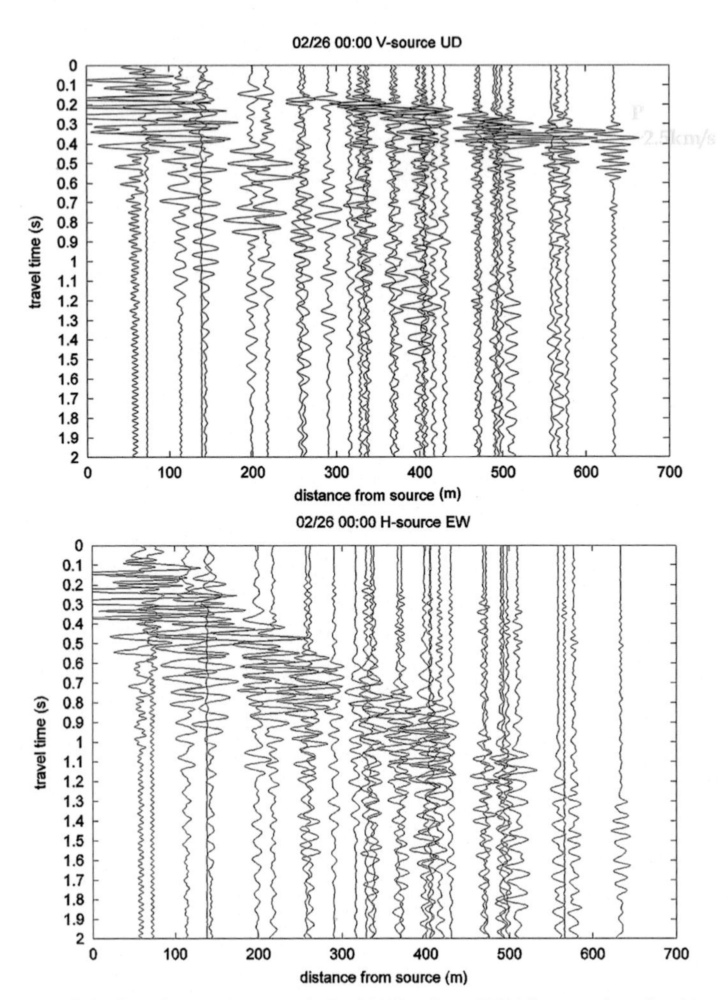

Figure 7.3 Calculated seismic records for UD (top) and EW (bottom) excited by vertical and horizontal vibrations, respectively. Vertical axis: travel time (s). Horizontal axis: distance (m) from the source.

changes in these two stations were very large, the changes were clearly seen in the original waveforms. The travel time delay in P arrival was observed at station #6, but it was not large at station #7.

The changes are large for later phases rather than P and S first arrivals.

The travel time change due to the air injection was 12 ms after one day at station #6 (Fig. 7.7). We cannot see the clear travel-time change at station #18 (Fig. 8.3).

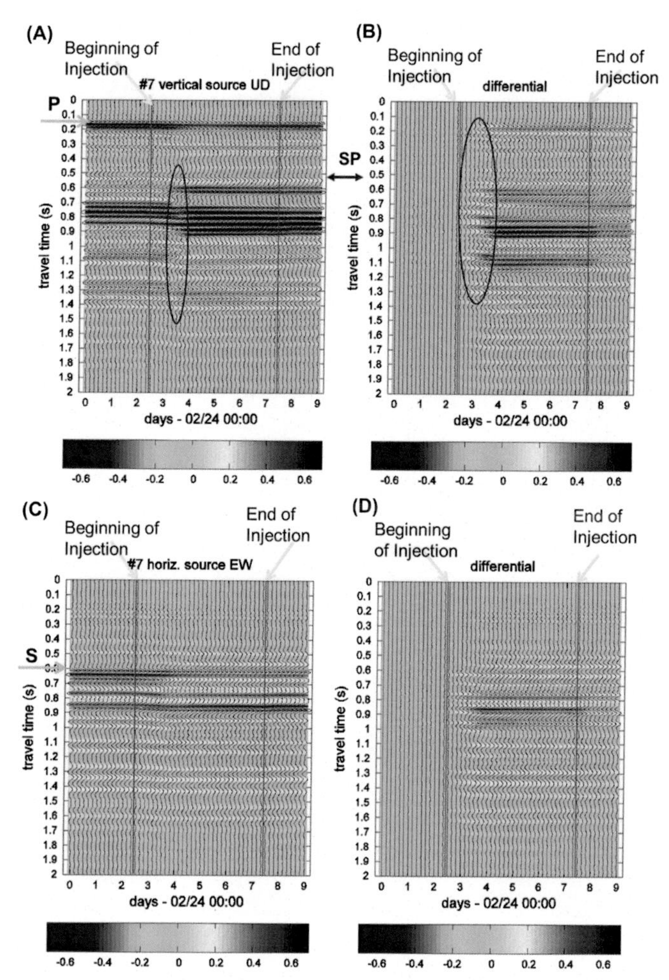

Figure 7.4 Seismic records at station #7. A, B and C, D are UD component records excited by vertical force and horizontal force, respectively. Original (A and C) and residual (B and D) to February 24 at 0:00 records. Vertical axis is travel time and horizontal axis is elapsed time from February 24, 2011. Scale at the bottom shows amplitudes of seismograms. Waveforms are shown for every 4 h. Two vertical lines in each diagram are the start and the end of the air injection.

7.2.4 Imaging by Use of Residual Observed Records

By use of reverse-time method (see Chapter 4, Kasahara et al., 2011a), the residual waveforms were backpropagated. The results are shown in Fig. 7.8 (Kasahara et al., 2013c). They used *P* portion of waveforms.

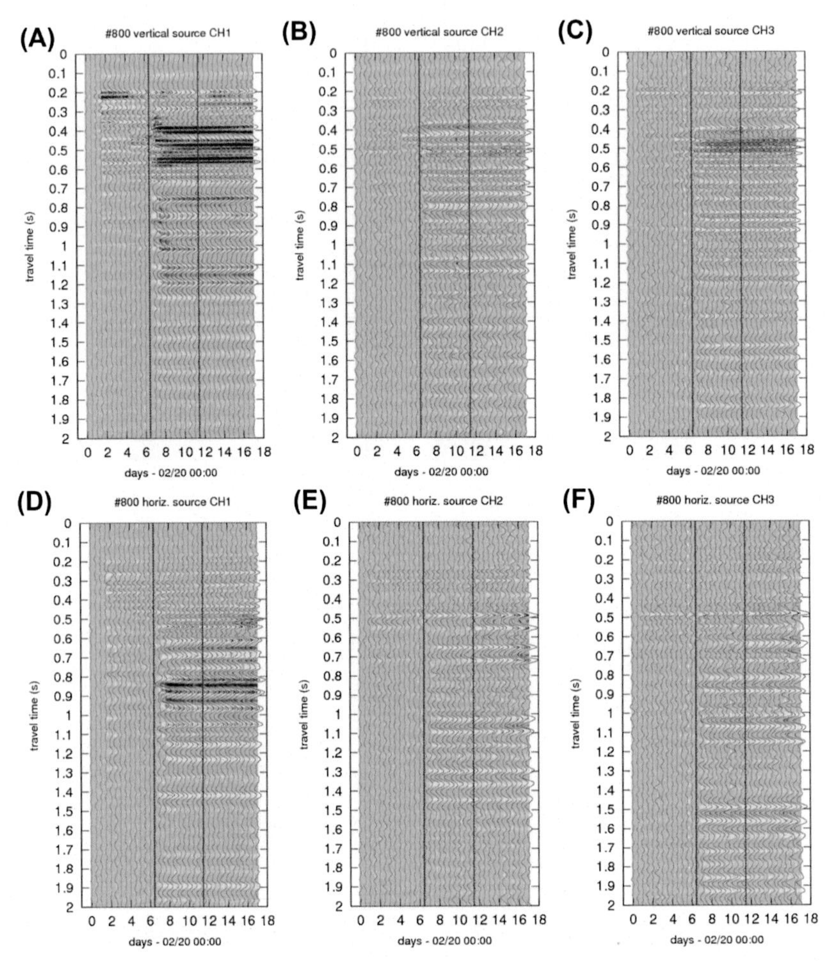

Figure 7.5 800-m borehole residual records. A–C are UD, NS, and EW components for vertical force and D–F are UD, NS, and EW for horizontal force.

According the elapsed time, the disturbed zone moves from west to east. Although air was injected into the 100 m depth, the most intensively disturbed zone was concentrated on the shallow depth. However, the boundary of affected zone is very sharp.

The same observations are identified at each station shown in Fig. 7.9. Most largely affected stations are distributed from the injected well to the eastward. The 800-m borehole is located at the extension of this eastward affected zone suggesting the presence of fractures down to 800 m depth.

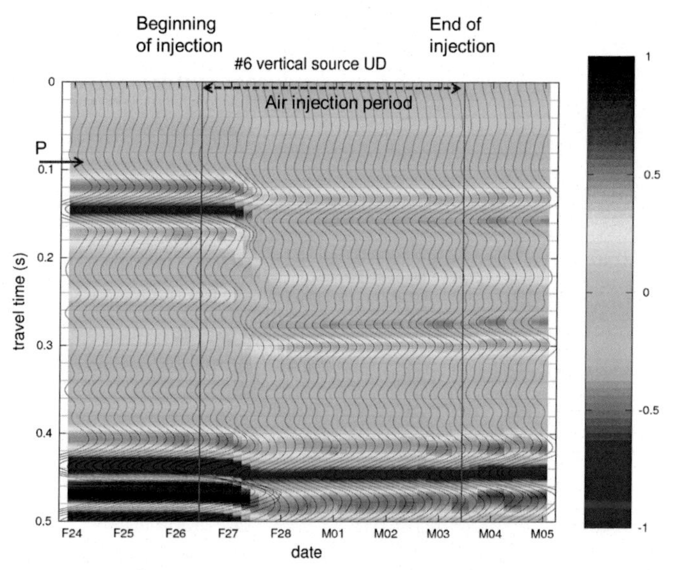

Figure 7.6 Evidence of observed waveform changes caused by air injection started at 12:00 local time on February 26 and ended at noon on March 3, 2011. The vertical axis shows the arrival times, and the horizontal axis shows the date since February 24. The dramatic changes caused by air injection beginning on February 27 can be clearly identified. *(After Kasahara, J., Ito, S., Fujiwara, T., Hasada, Y., Tsuruga, K., Ikuta, R., Fujii, N., Yamaoka, K., Ito, K., Nishigami, K., 2013c. Real time imaging of CO_2 storage zone by very accurate stable-long term seismic source, Energy Procedia, 37, 4085–4092.)*

All residual seismic records show very large change of waveforms on later arrivals of P and S. The interpretation is not easy because of heterogeneous underground structure. The synthetic seismograms using layer structure cannot explain such complicated waveforms. The large arrivals after P and S first arrivals seem as the diffracted phases or P-to-S or S-to-P converted phases. It is necessary to use heterogeneous V_P and V_S structure for the waveform computation.

Even though the interpretation of waveform changes is not easy to explain, the time reversal of the P-wave portions showed the clear tendency of eastward expansion of affected zone according to the elapsed time. Although the behavior of air is not the same as supercritical CO_2, the first step for the CCS and CO_2-EOR can be obtained by the combination of seismic ACROSS and multireceivers.

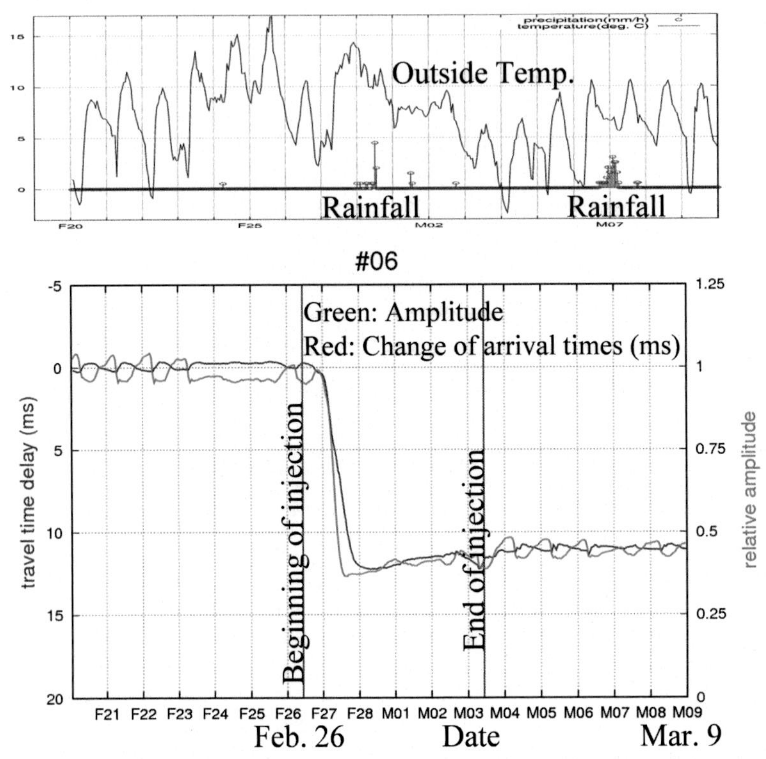

Figure 7.7 P travel-time delays and amplitude changes with elapsed days for station #6. P travel-time delays at station #6 decreased 12 ms after one day of air injection, but ones at #18 did not change. Amplitudes show similar tendencies. However, two rainfalls change amplitudes at station #18, but this is not seen in station #6. Daily variations of amplitudes identified at #6.

7.3 TIME-LAPSE EXPERIMENT USING MODIFIED CONVENTIONAL SEISMIC SOURCE TO EVALUATE THE NEAR-SURFACE EFFECTS

The near-surface effects seem to bring large temporal variation on the observed data. Kasahara et al. (2014a, b, 2015a, b) evaluate the usefulness of borehole geophones and influence of rainfall and temperature variation through the time-lapse study in a green tuff region in Japan. In this section, we describe the result of field experiments at this region by use of improved conventional electromagnetic seismic vibrator (Kasahara et al., 2014b). Two boreholes were drilled into the green tuff layers down to 200 m depth and cores

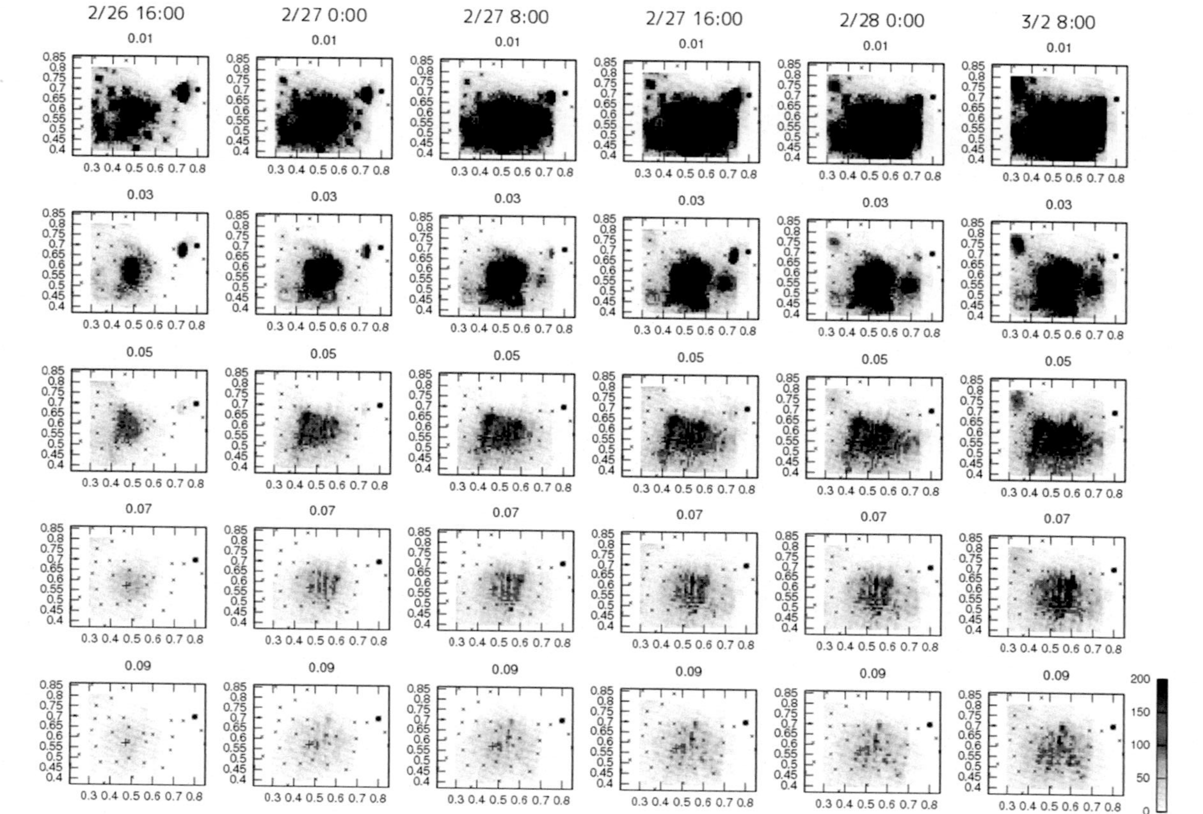

Figure. 7.8 Time lapse of imaging result considered to be air–water migration caused by air injection. (Top to bottom): imaging results of depth slice from 100 to 900 m. (Left to right): time slice from 16:00, February 26 to 8:00, March 2. Each small image shows 450 m (north–south) × 600 m (east–west). The air injection was conducted from February 26 to March 3, 2011. *(After Kasahara, J., Ito, S., Fujiwara, T., Hasada, Y., Tsuruga, K., Ikuta, R., Fujii, N., Yamaoka, K., Ito, K., Nishigami, K., 2013c. Real time imaging of CO_2 storage zone by very accurate stable-long term seismic source, Energy Procedia, 37, 4085–4092.)*

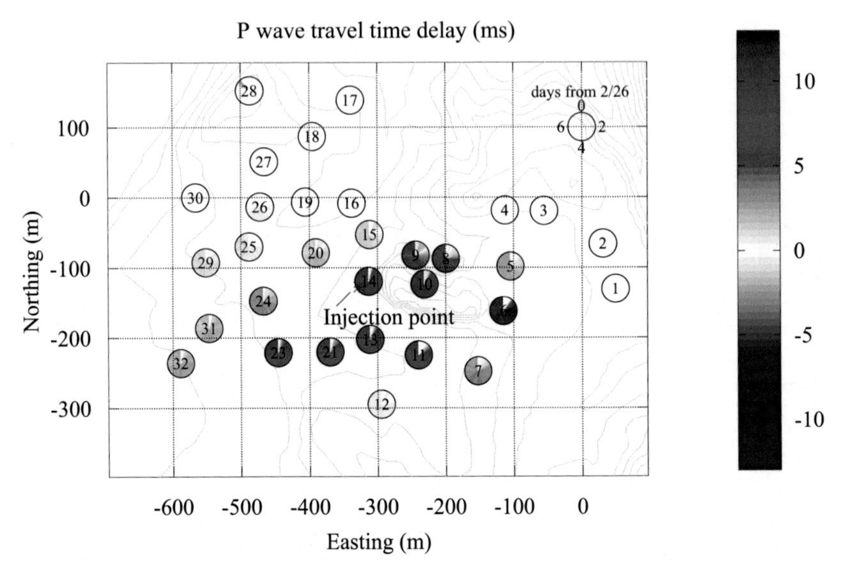

Figure 7.9 P first arrival delay at each station. Each circle shows delay time during nine days from February 26 by clockwise.

were obtained for the entire depth. Three-component borehole geophones were placed at depths of 70 and 200 m. Using these borehole geophones, zero-offset vertical seismic profile (VSP) and a walkaway VSP were carried out to obtain the shallow structure of the test field down to 200 m. After the drilling and VSP experiments, the time-lapse experiments for a few days to a week were done intermittently from August 2013 to February 2014.

In July and August in 2014, the time-lapse measurements in the green tuff region were also done in the same place to keep the same setting to the vibrator to the ground (Kasahara et al., 2015a,b).

7.3.1 Seismic Source for the Time-Lapse Experiment

In this time-lapse experiment an improved conventional electro-magnetic seismic vibrator was used. The vibrator was controlled by use of the accurate GPS time base (Kubota et al., 2014) (Fig. 7.10). The one sweep sequence was from 10 to 50 Hz during 100 s with 5-s rest. One-hour sweep sequence is comprised of 34 sweeps. The amplitude spectra of source and noise for feedback operation are shown in Fig. 7.11.

Figure 7.10 Improved electromagnetic vibrator used in this time-lapse experiment (Kubota et al., 2014).

7.3.2 Field Setting

The test site is in green tuff quarry area with 3 × 5.5 km in northeast Japan (Fig. 7.12). The locations of the seismic source and geophones are shown in Fig. 7.12. The zero-offset and walkaway VSP experiments and time-lapse experiments were carried out from August 2013 to February 2014. One hundred twenty-six geophones with the eigenfrequency of 14 Hz buried at 30 m depth were used for the time-lapse study. Two 3C borehole geophones with the eigenfrequency of 4.5 Hz, placed at the depths of 70 (KOM-1) and 200 (KOM-2) m, were also used for this experiment. The location of the borehole is shown in Fig. 7.12. Whole data observed by 126 buried geophones and 70 and 200 m 3C borehole geophone were digitized by 24 bit A/D with 1 kHz sampling. The GPS time base was used for whole study.

The zero-offset VSP and walkaway VSP were carried out using the borehole with KOM-1 and KOM-2. The length of walkaway was 430 m. Eight sweeps were repeated at every 10 m locations. The method of processing was similar to the ACROSS one (Nagao et al., 2010; Kasahara et al., 2010a). For the VSP surveys, 7 s sweep from 20 to 180 Hz was used.

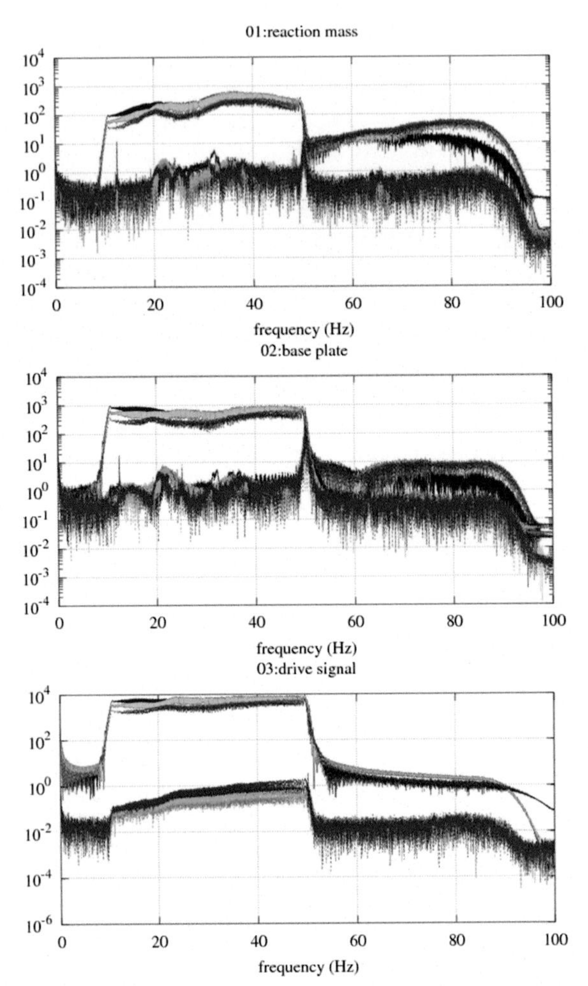

Figure 7.11 Source spectra and noise spectra of the system. From top to bottom: reaction mass, baseplate, and drive signal (Kubota et al., 2014). Vertical axis is output numbers from each sensor and horizontal axis is frequency in Hz. Each graph has signal and noise spectra.

7.3.3 Additional Data

7.3.3.1 Borehole Cores

The near-surface effects and rock type might control the time lapse. Three kinds of geophone setting at surface, 70, and 200 m depths were used to evaluate the near-surface effects on the time-lapse study. The cores obtained by drilling might help to consider the time lapse.

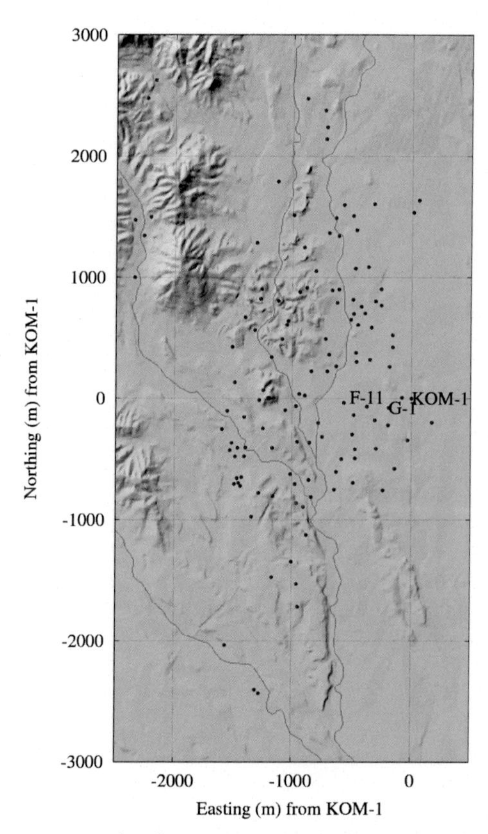

Figure 7.12 Seismic array used in this experiment. The borehole with geophones KOM-1 and KOM-2 is located at (0, 0). The source locations in this experiment were near G12 (350 m south of the borehole) and near the borehole. Each 14-Hz geophone was buried to 30 m depth. Elevation map by Geospatial Information Authority of Japan (GSI) was used.

Borehole cores down to 200 m deep were examined. The rock type is so-called green tuff. The layer shallower than 16 m is weathered layer comprised of soft sediments. There were some cracks and pores in the rocks. The core inspection suggests weathered sedimentary layer above 16 m and very porous green tuff layer below 16 m.

7.3.3.2 Zero-Offset VSP

The interpretation of zero-offset VSP is shown in Fig. 7.13. The V_P and V_S of weathered layer above 16 m are very slow. The V_P of green tuff layer below 16 m is ~2000 m/s (right in Fig. 7.13). The maximum V_S is 230 m/s.

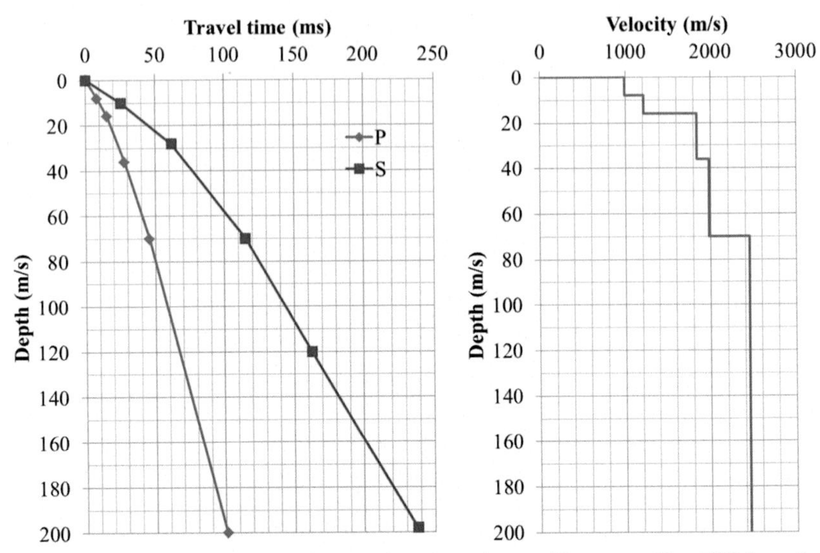

Figure 7.13 Travel time of V_P and V_S (left) in ms obtained by zero-offset VSP from 0 to 200 m. (Right) V_P structural model for the calculation of synthetic seismograms using zero-offset VSP data.

7.3.3.3 Walkaway VSP

The results of walkaway VSP and the transfer functions between the source (near G12) and boreholes (KOM-1 and KOM-2) are shown in Fig. 7.14. The distance origin is the location of the borehole. The source location of the transfer function during the time lapse was near G12. The transfer function between source (near G12) and KOM-1 (or KOM-2) is also shown in Fig. 7.14. It is noted that the frequency of the walkaway VSP and the transfer function of the time lapse are quite different. The frequency band of VSP and the time-lapse study were 20–180 Hz and 10–50 Hz, respectively. Although it is difficult to identify the P first arrivals and S arrivals in the surface geophone records, we can recognize clear P first arrivals on NS and EW components of 70-m borehole geophones and vertical and EW components of 200-m borehole geophones. S phase is not clear for all records, but very large phase arrived after S phase.

We calculated synthetic seismograms using the velocity model shown in Fig. 7.13 assuming 50-Hz zero-phase Ricker wavelet as the source wavelet. The results are compared to the observation (Fig.7.15).

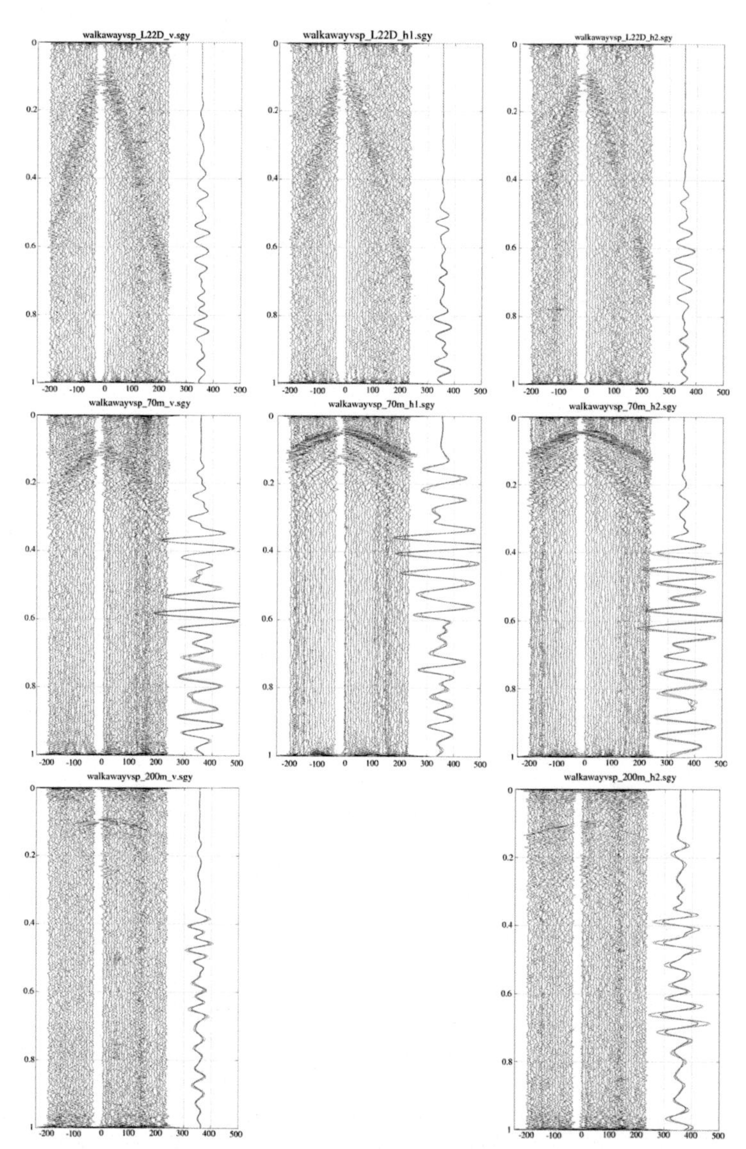

Figure 7.14 Results of walkaway VSP seismograms (black) and transfer functions (red) at (top) surface geophone, (middle) 70-m-deep geophone, and (bottom) 200-m-deep geophone in the borehole. The transfer functions are from the source near G12 to the borehole. The borehole records are placed at corresponding distance from the source. From left to right are records of vertical, NS, and EW components.

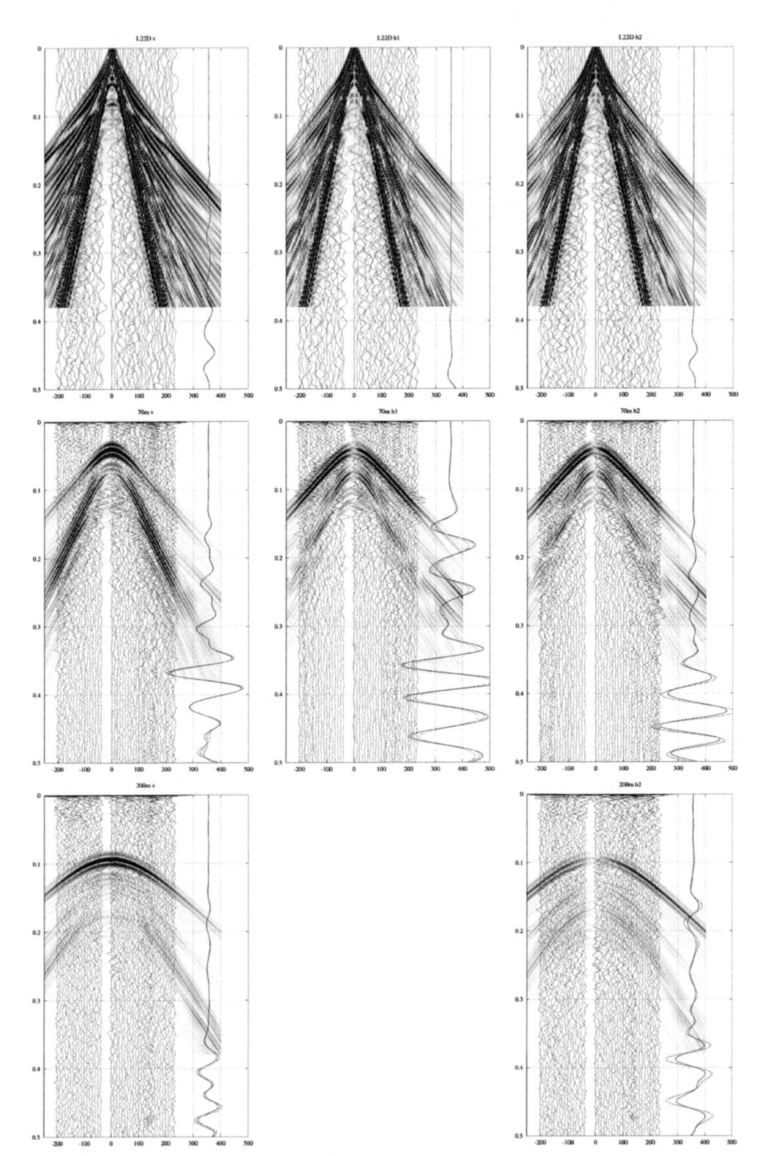

Figure 7.15 Comparison of walkaway seismograms (black), transfer functions (red) at (top) surface geophone, (middle) 70-m-deep geophone, and (bottom) 200-m-deep geophone in the borehole and synthetic seismograms using model shown in Fig. 7.13.

The synthetic seismograms at surface show very large surface wave arrivals. The observation of walkaway VSP obtained by surface seismometers indicates large arrivals by surface waves, but there are no apparent surface wave arrivals on records of 70- and 200-m-deep borehole seismometers. The near-G12 vibration between 10 and 50 Hz has clear arrivals for P and S on observed records of70- and 200-m-deep borehole seismometers.

7.3.4 Result of the Time-Lapse Experiments at Green Tuff Region

Twelve hours data for each day were stacked in frequency domain. The 100 H_{vV} (vertical vibration and vertical geophone) source-gather transfer functions on July 18 and July 23 is shown in Fig. 7.16. It is difficult to recognize any distinct changes among five days' records except background noise.

Fig. 7.17a shows details of waveforms each day for a five-day period in 15 stations out of 126 geophones. The distance of 15 stations are from 11 to 575 m away from the source. As we see the same results shown in Figs. 7.16 and 7.17a, we cannot recognize distinct changes for five days in these 15 stations. However, the residual waveforms show the changes of waveforms that occurred at some stations as seen in Fig. 7.17b. The repeatability of this vibrator is not perfect and the residuals increase day by day. The largest amplitude changes occurred at the station F11 at 451 m offset distance. We also notice smaller amplitude changes at later arrivals in some stations.

The results for 200-m borehole geophone and F11 vertical geophone buried at 30 m depth in December (Fig. 7.18) are compared. The shown traces are the transfer functions from the source near the borehole with KOM-1 and KOM-2 to the geophones. The borehole seismic traces at 200 m depth indicate that a very large temporal change occurred at 0.8 s arrival time in spite of the receiver depth. The F11 traces at 30 m depth indicate waveform changes at 0.37 s and after 0.7 s. Because the distance to the borehole geophone is only 200 m, the apparent average speed of the phase at 0.8 s arrival is c.a. 250 m/s. The changes of the arrivals at 0.8 s seem

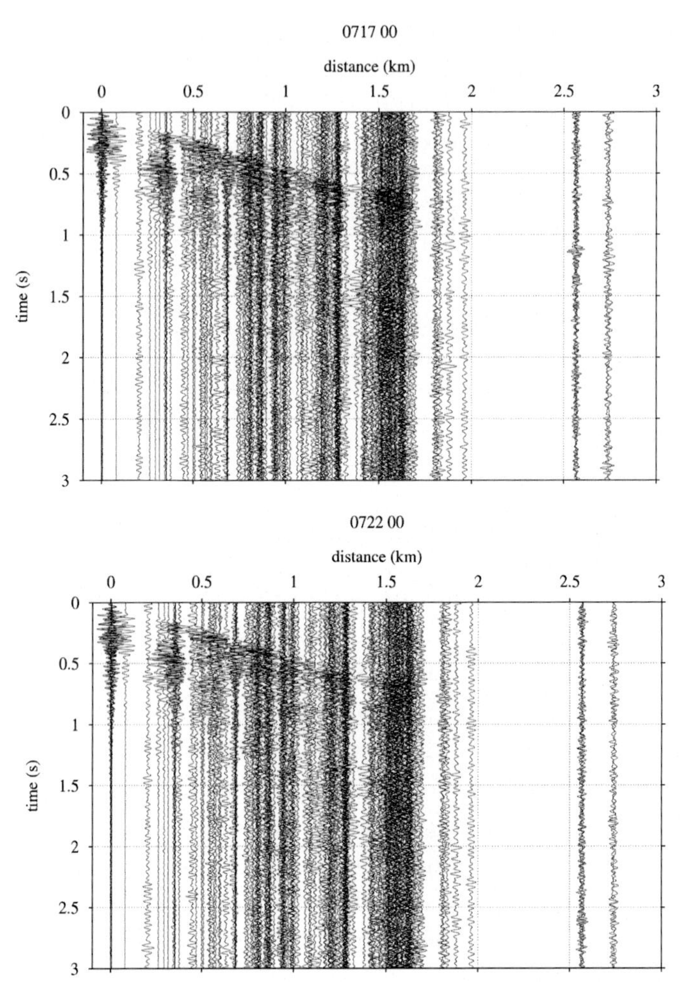

Figure 7.16 Comparison of the 100 H_{VV} transfer functions from July 18 and July 23. The horizontal axis is source-receiver offset distance in kilometers and the vertical axis is travel time in seconds. There are some noisy traces on August 18.

to be caused by the outside temperature change for the surface wave traveling through a very shallow part or Stoneley (tube) waves.

$H_{\nu V}$ (gray) and residual $H_{\nu V}$ (black) are shown in Fig. 7.19. The dashed line is the velocity of 300 m/s and the large-amplitude phase appeared at 450 m offset distance and traveled by approximately 300 m/s. This large phase appeared on August 16, 2013. One possibility of this phase is sound waves traveling in the subsurface cavity, which was an abandoned former quarry mine.

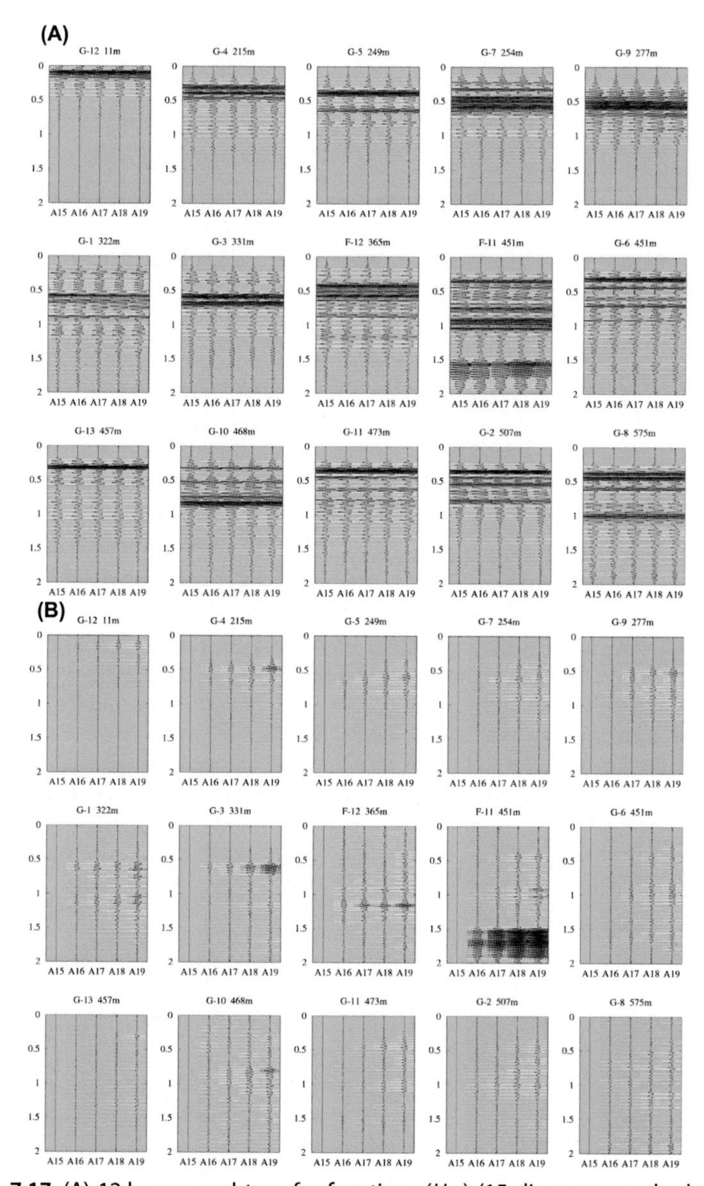

Figure 7.17 (A) 12 h averaged transfer functions (H_{VV}) (15 diagrams on the left side) and (B) residual transfer functions (15 diagrams on right side) for 5 days in 15 stations. The vertical axis is travel time in seconds, and the horizontal axis is an elapsed day. Source location was at the source (near G12).

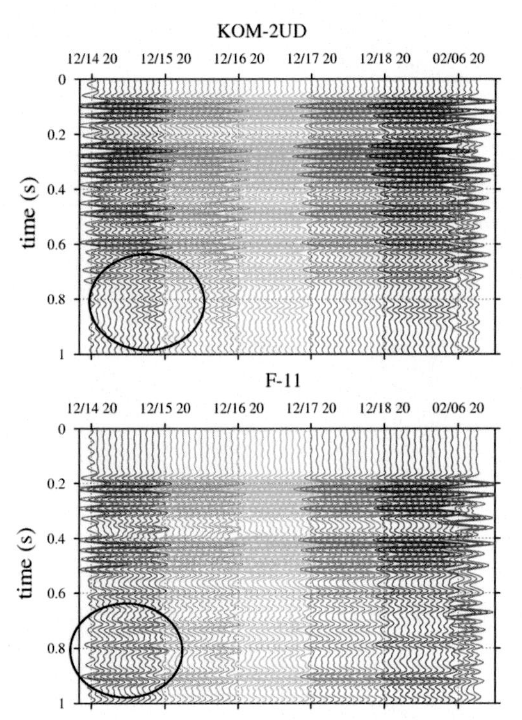

Figure 7.18 Time-lapse records of 200-m borehole seismometer and F11 (buried at 30 m depth with 378 m offset distance) from December 14 to 18, 2013, during 8 p.m. and 8 a.m. and 8–12 p.m. of February 6, 2014. Both records are vertical components. The source location was near the borehole with KOM-1 and KOM-2. Each trace is hourly transfer function. Vertical axis is travel time and horizontal axis is aligned every hour during the nighttime of each day. The circles show portions having most distinct waveform changes.

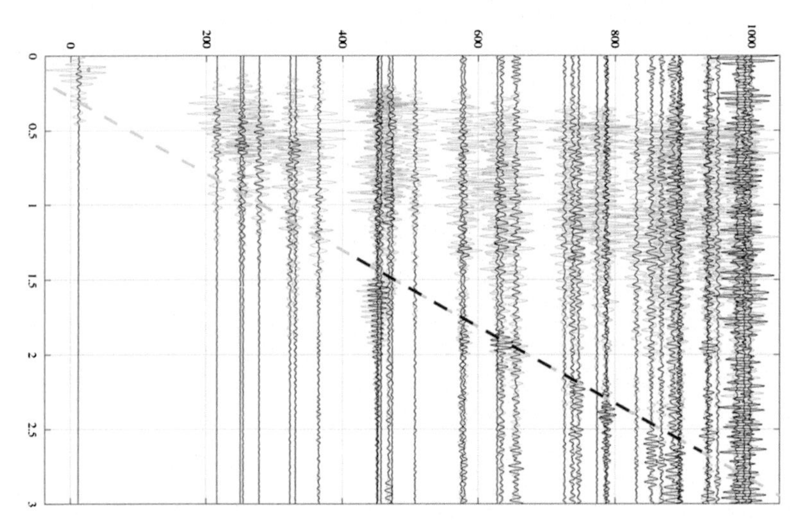

Figure 7.19 Comparison of residual H_{vV} transfer function (*black* wiggles) and original H_{vV} transfer functions (*gray* wiggles). The vertical axis is travel time in seconds and horizontal axis is offset distance in meters. Dashed line is $V = 300$ m/s corresponding to the phase that appeared at station F11 after August 16, 2014.

The comparison of December 2013, July 2014, and August 2014 is shown in Fig. 7.20. The source locations in 2013 and 2014 are slightly different. The water level changes of abandoned green tuff quarry cavities and precipitation are also shown in the same figures. As seen in Fig. 7.20, the first arrival of P seems to stay the same, but the later phase of P varies on amplitudes and arrival times. The precipitation and the water levels due to rainfalls affected the 70-m borehole (KOM-1UD) waveforms for later than 0.2 s and 80 m distance one for 0.4 s. The 200 m borehole (KOM-2UD) and 346 m distance ones seem less sensitive. The reason could be the depth of water table and distance to the underground water pool.

Using cross-correlation of P part, the travel-time change is measured (Fig. 7.21). The residual waveforms are backpropagated, and the results show the eastward direction as the origin of travel-time change. The scattered wave field concentrates in the eastward direction suggesting the cause of travel-time change is eastward.

7.3.5 Discussion in This Section

The comparison of walkaway records at the surface and borehole geophones shows very large surface wave arrivals by surface geophones and small amplitude for surface wave portions by 70 and 200 m borehole. The borehole records show clear P and S arrivals. This suggests that the shallow seismometer records were strongly affected by surface waves and borehole records were less influenced by surface waves.

The temporal changes were very small on P and S first onsets, but on the later arrivals of P and S were very large. Especially during the night in December, we can recognize the distinct waveform changes at very late arrival phases. The average speed of arrivals at 0.8 s seen in the 200-m borehole is c.a. 250 m/s and they were strongly affected by the outside temperature change, probably frozen of near-surface soils.

Distinct changes on transfer functions were observed at F11, which is 451 m away from the source. The portion of changes is

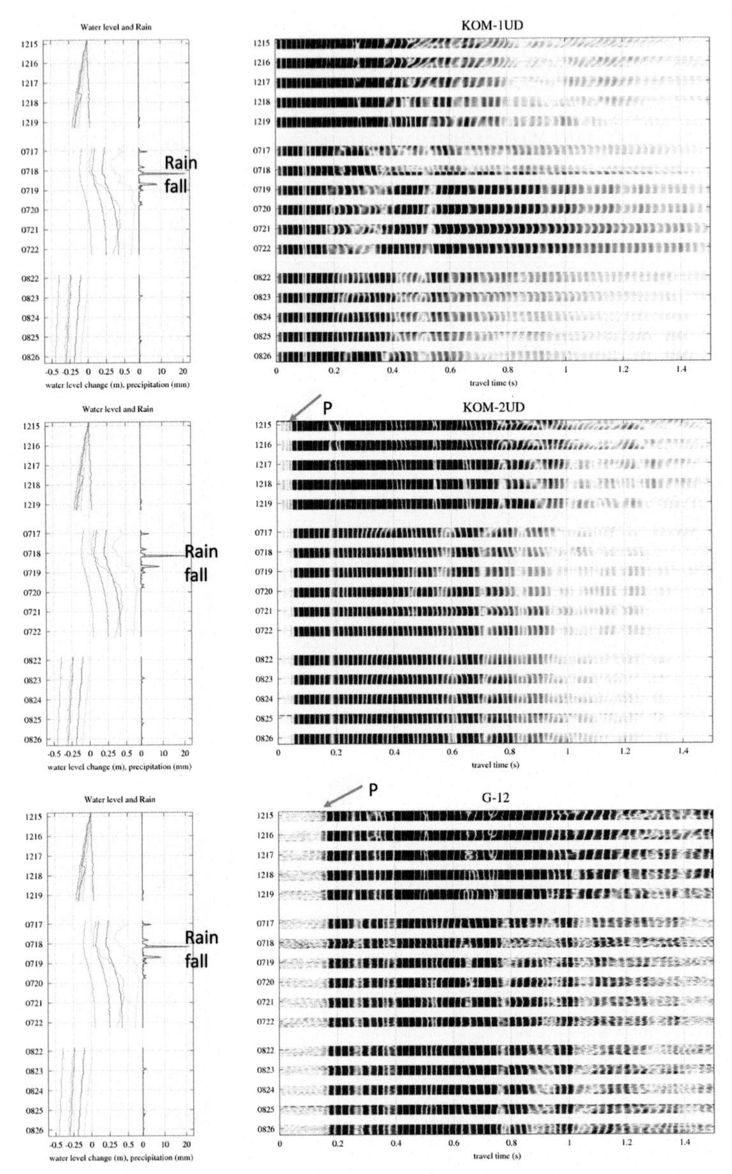

Figure 7.20 (Right) Time-lapse of three stations in December 15–19, 2013, July 17–22, 2014, and August 22–26, 2014. From top to bottom: borehole at 70 m depth (KOM-1UD), borehole at 200 m depth (KOM-2UD), 80 m distance at 30 m buried, and 346 m distance at 30 m buried. (Left) water level change (m) at underground abandoned mine cavities and precipitation (mm).

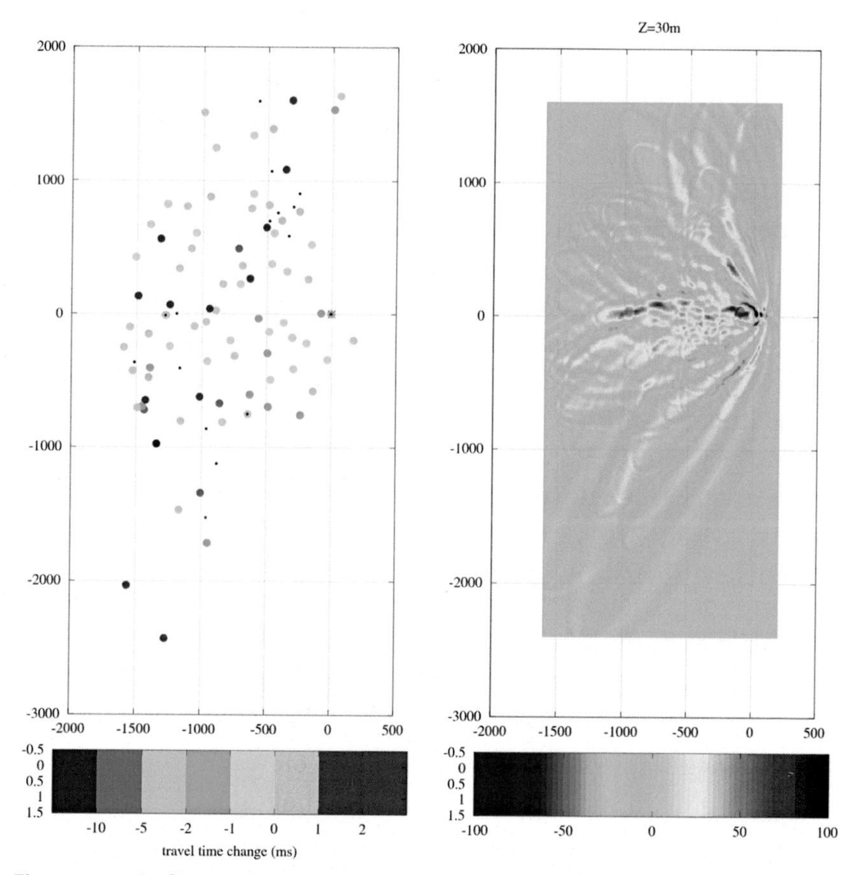

Figure 7.21 (Left) Travel-time changes of P arrivals in July and August, 2014, observed in the green tuff region in Japan. The travel-time changes were obtained by cross-correlation of P phase with correlation greater than 90%. (Right) Backpropagation of residual waveforms (Kasahara et al., 2014c). Horizontal direction is W to E and vertical direction is S to N. The scattered wave field concentrates in the eastward direction.

identified at very slow arrivals. It is thought that the cause of the changes is close to the station F11 and/or near surface. The velocity from the traveling is very slow, namely 300 m/s, suggesting traveling near surface. During the experiment in August, 2014, there was no rain and the air temperature was more than 30°C. The cause of change around the station F11 was not clear; it might be caused by the migration of aquifer at very shallow depth to induce the change of surface wave velocity.

7.3.6 Summary of Time-Lapse Experiment in Green Tuff Region

A time-lapse experiment was carried out in the green tuff region of 3 × 5.5 km in Japan. In this study, an improved conventional electromagnetic vibrator was used as seismic source. To evaluate the usefulness of borehole geophones in the time-lapse study, zero-offset VSP and walkaway VSP were also conducted.

The comparison of walkaway VSP records and fixed location record at the source (near G12) suggests that the 70- and 200-m borehole seismometers do not show large amplitudes for surface wave although the seismometer at surface has very large surface wave arrivals.

The transfer functions between the source and receivers were obtained by the division of observed waveforms by source signature in frequency domain. Using the subtraction of waveforms for each day with 12-h stacked data from the first day, residual waveforms were calculated.

Although the original transfer functions at KOM-1 and KOM-2 do not show distinct temporal changes on waveforms, the residual transfer functions indicate distinct change of waveforms for later arrivals for all 126 buried stations and borehole seismometers. The waveform changes of later phase at F11 are extremely large. The waveform change seems to occur near the station F11 and be radiated to other stations. One of this large changes seems to be associated with sound waves traveling the former quarry cavities in subsurface.

The time lapse during December 2013 shows large waveform changes even in borehole geophone at 200 m depth. The average velocity for this phase is 250 m/s; it could be tube wave traveling along the borehole or surface wave traveling near the surface. The temporal changes in whole stations were distinct during the midnight of observation suggesting possible cause by frozen of near-surface layer. This is discussed in the near-surface effect in Chapter 8.

7.4 FIELD TEST IN SAUDI ARABIA

In this section we show results of the time lapse in Al Wasse field in Saudi Arabia. The Al Wasse water-pumping site is approximately

120 km east of Riyadh and was selected as a trial site. The intention was to observe the changes in aquifers induced by pumping operations. One ACROSS source unit was installed there in December 2011, and a field test was conducted (Kasahara et al., 2015c). One and half months of data between December 19, 2012, and February 08, 2013, and four months of data in April–December 2015 were obtained (Kasahara et al., 2016a, b). The seismic records from the excitation by the ACROSS seismic source were analyzed. In addition, cross-correlated background noise during ACROSS operation was used for the seismic interferometry technique.

7.4.1 Al Wasse Test Site

The lithology of the Al Wasse field comprises limestone–dolomite, and unconsolidated sand and consolidated sandstone. The ground surface is locally covered with sand dunes. The water table is at the depth between 400 and 500 m deep in sandstone layers. Unfortunately, there are no publicly accessible seismic profiles in this area. This has made understanding the seismic phases difficult.

The ACROSS seismic source is at the center of an array of geophones shown in Fig. 7.22. Thirty-one 3C surface geophones with the eigenfrequency of 1 Hz deployed in a 3 × 2 km area spaced at 500 m were used (Fig. 7.22). The distance between the ACROSS source and each station was approximately 500–1800 m. The ACROSS seismic source was mounted in a concrete base 5 m (L) × 5 m (W) × 3 m thick and buried at 3 m deep. The source was operated as 10–50 Hz between December 2012 and February 2013, and 10–40 Hz after March 2015. The length of one nonlinear sweep was 200 s. In an hour, there were 16 sweeps and 400 s sleep time. The source spectrum observed in the source room is in Fig. 7.23 and the noise spectrum was calculated (see Appendix C.2). The S/N in the ACROSS room was more than 100 even though the geophone was just next to the ACROSS seismic source.

The three-component geophone data at each grid station (Fig. 7.22) were continuously recorded by a 24-bit data logger with the GPS time standard. The sampling rate was 200 Hz, but resolution of temporal variation is 0.2 ms because the data can be resampled to

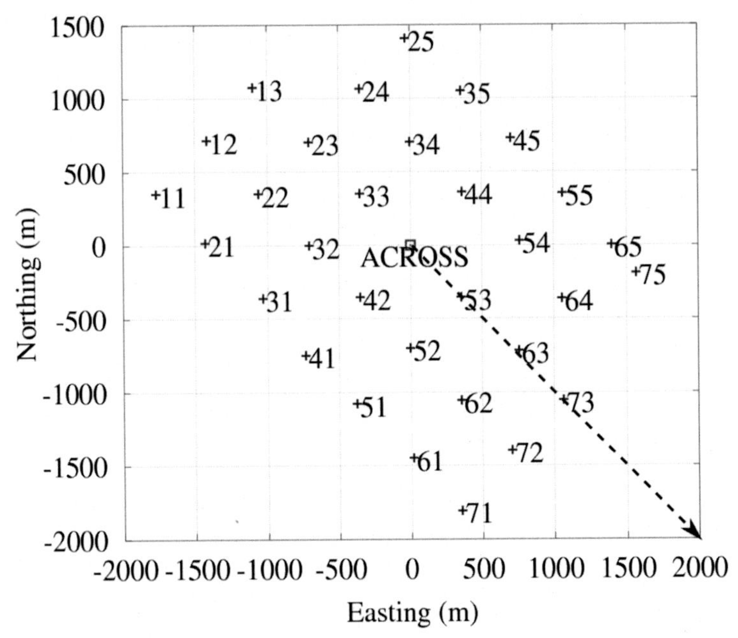

Figure 7.22 Location map of seismic ACROSS source at (0, 0) and 31 3C-1 Hz geophones. The scale is in meters. The rotation axis of the motor is NE—SW. Refraction survey using the ACROSS seismic source was conducted in December 2015 along the dotted line with arrow.

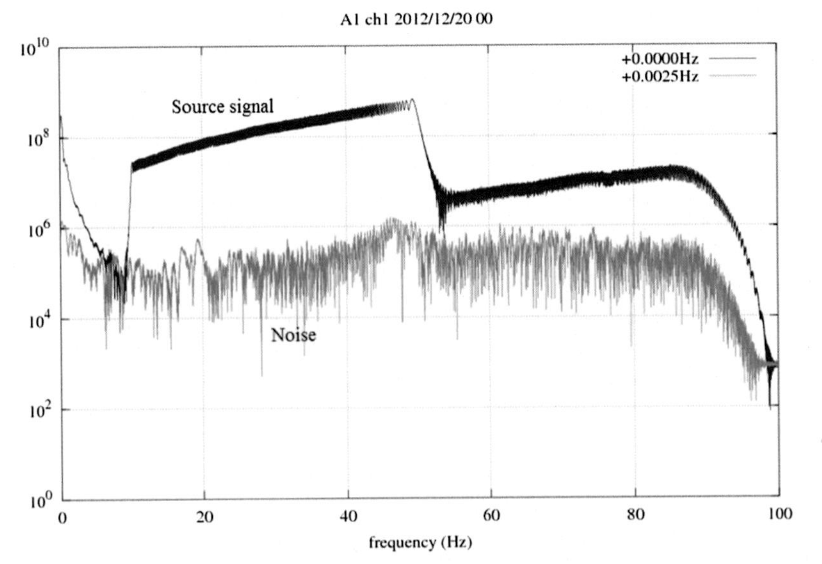

Figure 7.23 Source and noise spectrum of ACROSS in Al Wasse obtained by a vertical geophone at 1 m from source. The method of separation is described in Appendix C.2. The S/N is more than 100 even though the geophone is just 1 m from the source.

get enough time resolution. The geophones were placed at the ground surface, which might bring some degradation in repeatability as discussed in Chapter 9.

7.4.2 Data Processing

The data from December 2012 to February 2013 and from April to December 2015 were analyzed. The outline of the ACROSS processing is described in Section 3.3. One segment is 200 s and four segments of each geophone record were transformed from time domain to the frequency domain. The segments in each hour were stacked in frequency domain and divided by the source spectrum. The result was transformed to time domain.

7.4.3 Results Obtained With Active Seismic Source

The transfer functions between the ACROSS and each geophone are shown in Fig. 7.24. We obtained $V_P = 3.4$ km/s and $V_S = 1.9$ km/s for refracted P and S, respectively. P-phase was the first arrival identified in the H_{vV} transfer function. S-phase was interpreted as the fastest arrival after P according to the H_{rH} transfer function. Although a large phase after S was seen in H_{vV} and H_{rH}, the nature of this phase is not clear because our geophone stations were too sparse to determine the velocity structure from these records.

7.4.4 Results of Refraction Study

Because of no seismic reflection data in the test site in the past and geophone spacing was too wide to interpret observed time-lapse data, a seismic refraction survey was conducted using the ACROSS as the seismic source (Kasahara et al., 2016a, b). The line was in the middle of grid area (Fig. 7.22). Geophones were placed at every 50 m. The data at each station were recorded for 24 h. H_{vV} (vertical component excited by vertical force) and H_{hH} (horizontal geophone excited by horizontal force) were calculated. Fig. 7.25 is the result using 2 h from midnight in the local time. There are three distinct characteristics in the refraction records. Firstly, $V_P = 3.4$ km/s first arrivals disappears at distance further than 0.7 km, and, secondly, $V_P = 4.5$ km/s arrivals appears at the distance further than 1.4 km. As the third characteristic, there are no distinct reflections after

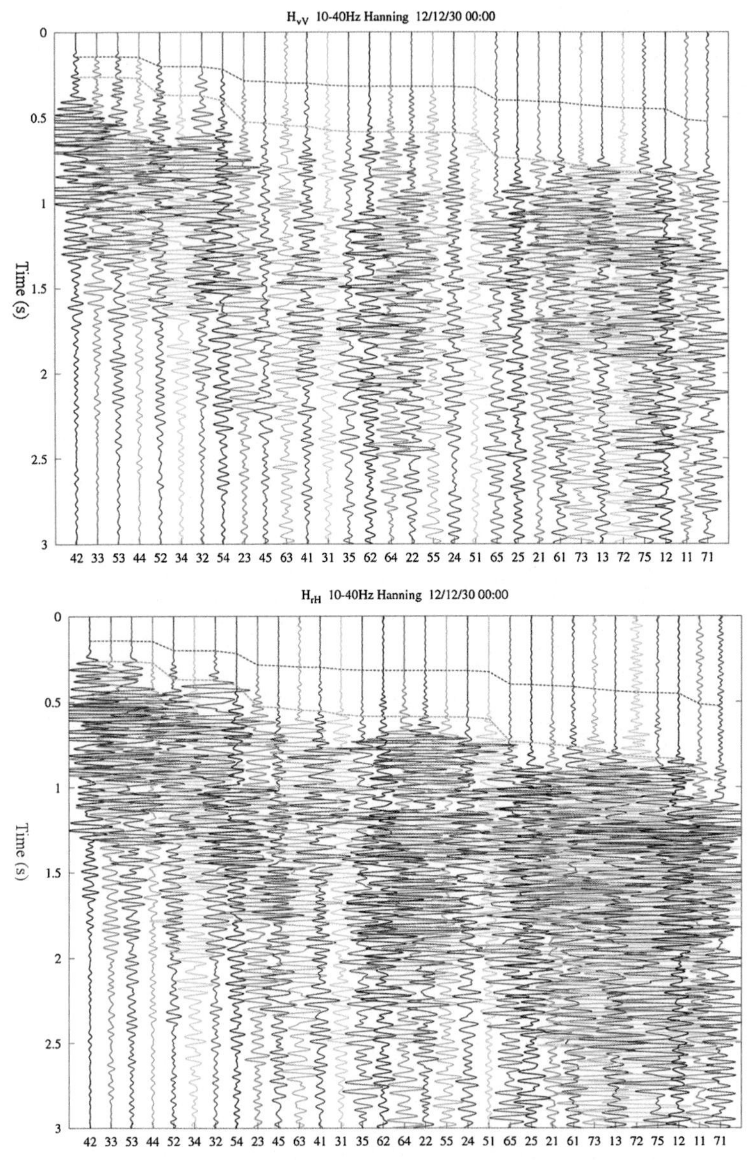

Figure 7.24 Transfer functions at 31 stations arranged by offset distances obtained from 2 h data: (H_{vV}, left) vertical geophone records excited by vertical single force; (H_{rH}, right) radial geophone records excited by horizontal single force. The 10–40 Hz Hanning window was used. The vertical axis represents the travel time in seconds. The horizontal axis is stations IDs, but is not proportional to the offset distance because multiple stations were at the same distance. Amplitudes were corrected by multiplying the squares of the offset distances. The horizontal upper and lower *dotted lines* represent $V_P = 3.5$ km/s and $V_S = 1.9$ km/s, respectively. The phase with very large amplitudes just after the arrival of the S-wave was not interpreted.

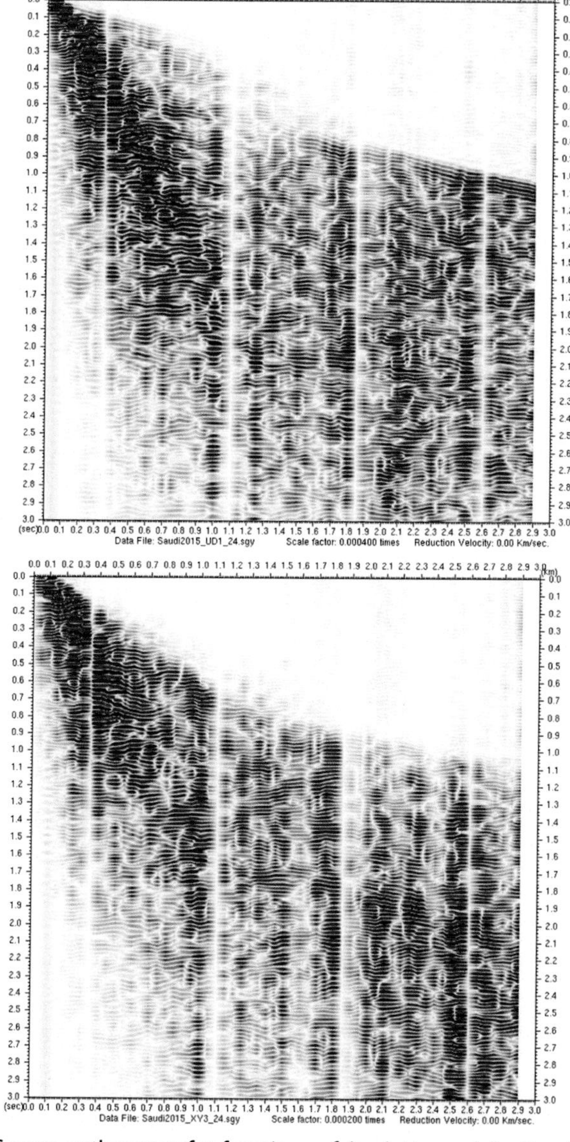

Figure 7.25 Source gather transfer functions of (top) H_{vV} and (bottom) H_{hH}. Vertical axis is travel time in seconds and horizontal axis is distance in km. Receiver spacing is 50 m. Source is ACROSS with 10–40 Hz frequency band. Two hours of data from 0 to 2 h were stacked (Kasahara et al., 2016a).

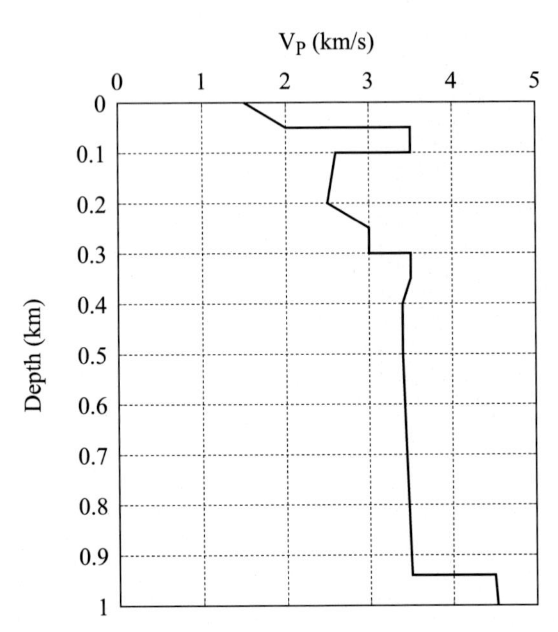

Figure 7.26 One of velocity structural models (model 0215C) to interpret the observed refraction records (Kasahara et al., 2016a).

4.5 km/s arrivals. In addition, there is no distinct phase between the 3.4 km/s and 4.5 km/s arrivals. The first and the second characteristics suggest the presence of low-velocity layer below 3.4 km/s top layer and the basement layer with $V_P \sim 4.5$ km/s. In the H_{hH} records, $V_S = 1.9$ km/s can be identified.

To interpret observed waveforms in Fig. 7.25, they carried out ray tracing and calculation of synthetic seismograms. A velocity structure model (model 0215C) was presented (Fig. 7.26) (Kasahara et al., 2016a). They compared the observed and synthetic seismograms (Fig. 7.27) (Kasahara et al., 2016a). The synthetic seismogram can explain major characteristics seen in the observation as described herein. The model 0215C has low-velocity layer at the depth between 100 and 900 m. The velocity gradient is complicated.

7.4.5 Time-Lapse Records Observed in Al-Wasse Field

The time lapses of the transfer functions for stations #42, #33, and #22 showed noticeable waveform changes, especially in the later

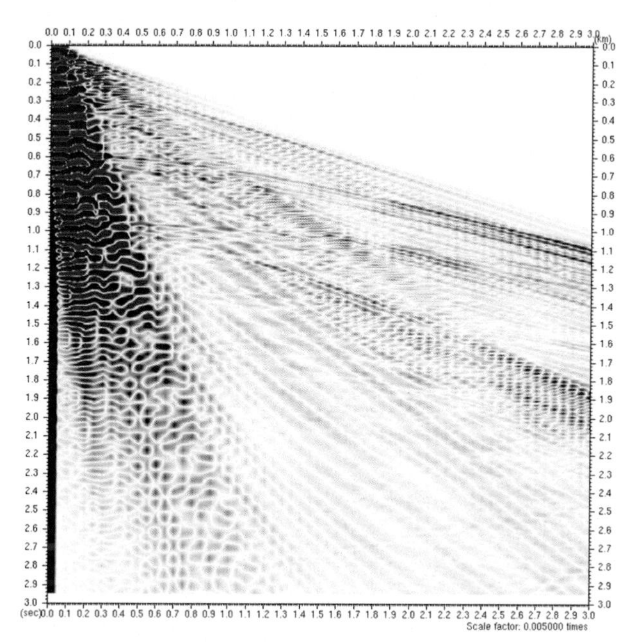

Figure 7.27 Synthetic (bottom) seismic records using the model 0215C. Waveform characteristics in the observed refraction records (Fig. 7.25) are well explained. Vertical axis is time in seconds and horizontal axis is distance in km (Kasahara et al., 2016a).

phases during the observed period (Fig. 7.28). Later arrivals showed very large temporal variation. The waveforms of later phases from January 20 to February 1, 2013, were most distinct. According to the interpretation of refraction records, the phases with large temporal change were surface waves. Because P arrivals are weak, it is difficult to see any variation of arrival time and amplitude from these records. When we look at the P arrivals (Fig. 7.29), we cannot see any significant variation of P arrivals during this period.

The comparison of P part of H_{vV} at station #34 during two years apart is shown in Fig. 7.30. The arrivals are almost the same for two datasets even if there are two years and two months between two periods. More details are presented in the Chapter 9. Including later arrivals, Fig. 7.31 shows time-lapse example of station #75. Because the offset distance of this station is 1.579 km, the arrivals at 0.8 s are the refracted phase form the basement layer. The amplitude around 0.8 s in 2015 is larger than those in 2012–2013. The 1.4 s phase in 2015 also had larger amplitudes than in 2012–2013. This temporal

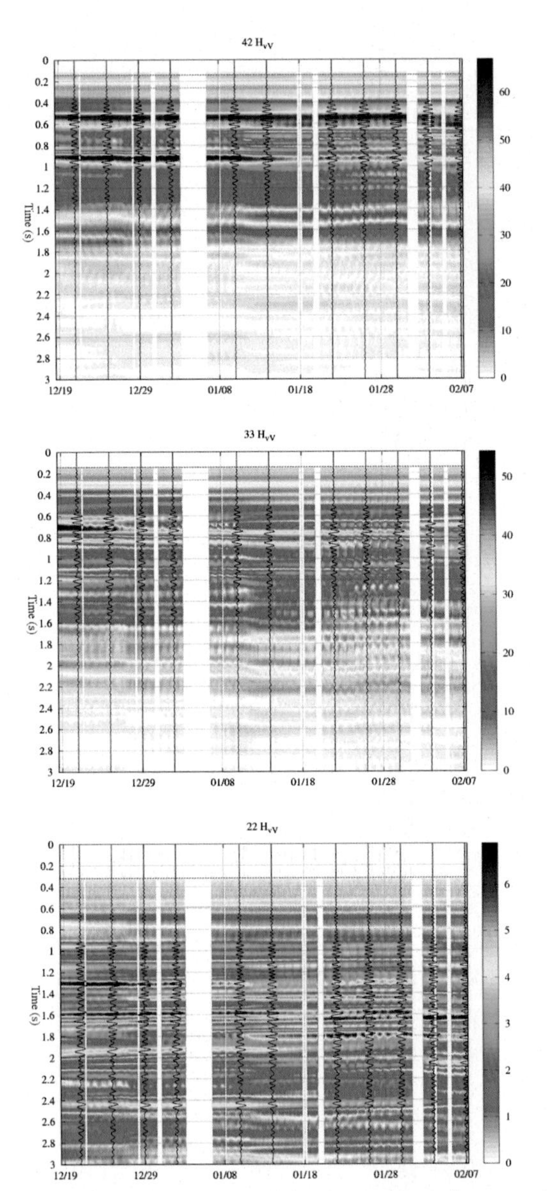

Figure 7.28 Seismic records (H_{VV}) observed by vertical geophone excited by vertical vibration for #42, #33, and #22. The offset distances of stations #42 (top), #33 (middle), and #22 (bottom) were 463 m, 530 m, and 1129 m, respectively. Vertical axis is travel time and horizontal axis is time from December 19, 2012, to February 07, 2013. Wiggles are waveforms, and right-hand side is the scale for amplitudes. Horizontal upper line shows $V_P = 3.5$ km/s line, and the second horizontal line is $V_S = 1.9$ km/s line.

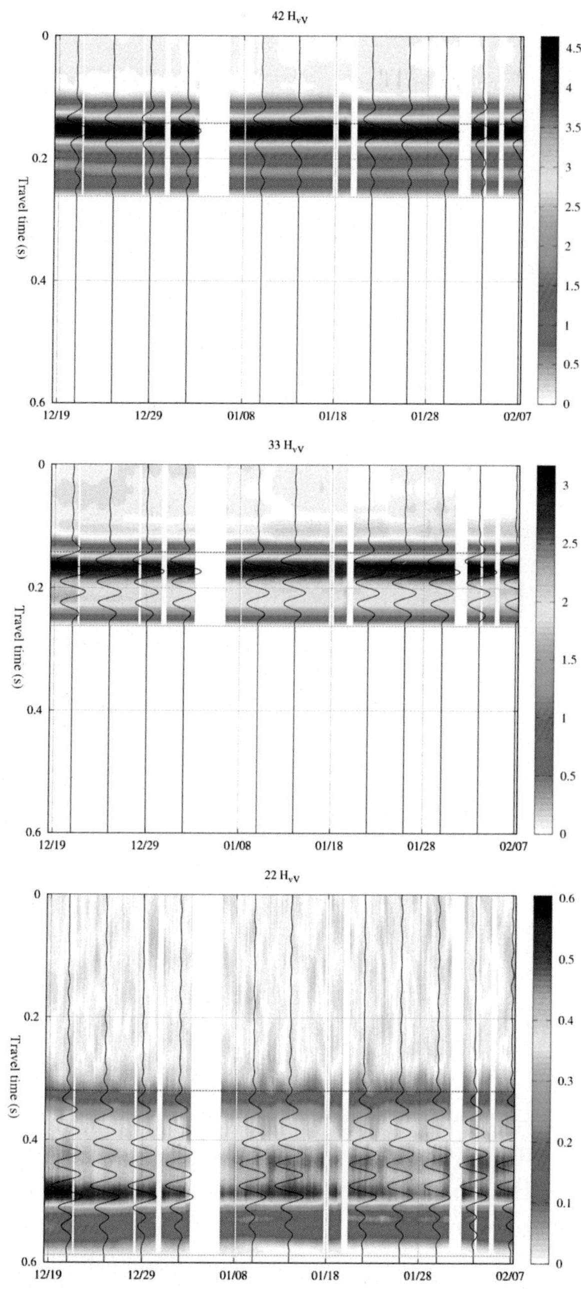

Figure 7.29 Time-lapse of H_{vV} transfer functions at three stations observed with vertical geophones excited by vertical force using 2 h data from December 18, 2012, to February 7, 2013. The offset distances of stations #42 (top), #33 (middle), and #22 (bottom) were 463 m, 530 m, and 1129 m, respectively. The upper and lower horizontal lines for each diagram represent 3.5 km/s for the P-wave and 1.9 km/s for the S-wave, respectively.

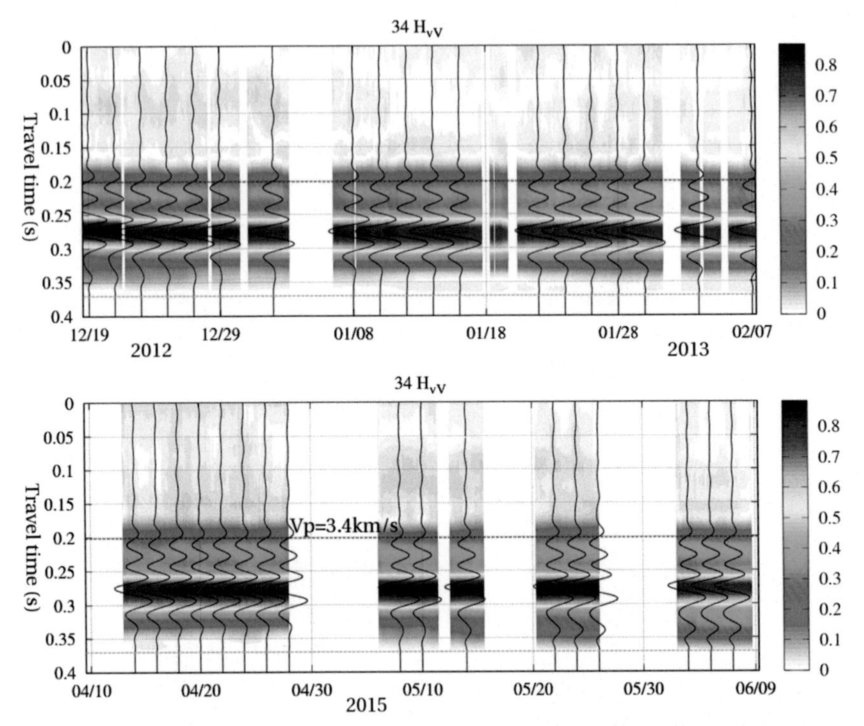

Figure 7.30 Time lapse of station #34 in (top) December 19, 2012, to February 7, 2013; in (bottom) April 15 to June 9, 2015. $V_P = 3.4$ km/s phase traveled through the top limestone layer. We cannot see noticeable change between two periods.

variation seen in wiggles of station #75 implies the effects caused by water pumping from aquifers.

7.4.6 Seismic Interferometry With Simultaneous ACROSS Operation

In this section we show one of the results using seismic interferometry of noise component (see Chapter 5). Kasahara et al. (2014c) attempted seismic interferometry by using the background noise during the ACROSS operation. The signal of the ACROSS seismic source in the frequency domain is a set of discrete line spectra. This feature enabled us to use the background noise observed at gaps of the signal spectra for passive seismic analysis. The method of the separation of noise from the actively operated ACROSS signal is described in Appendix C.2 based on the ACROSS methodology. An

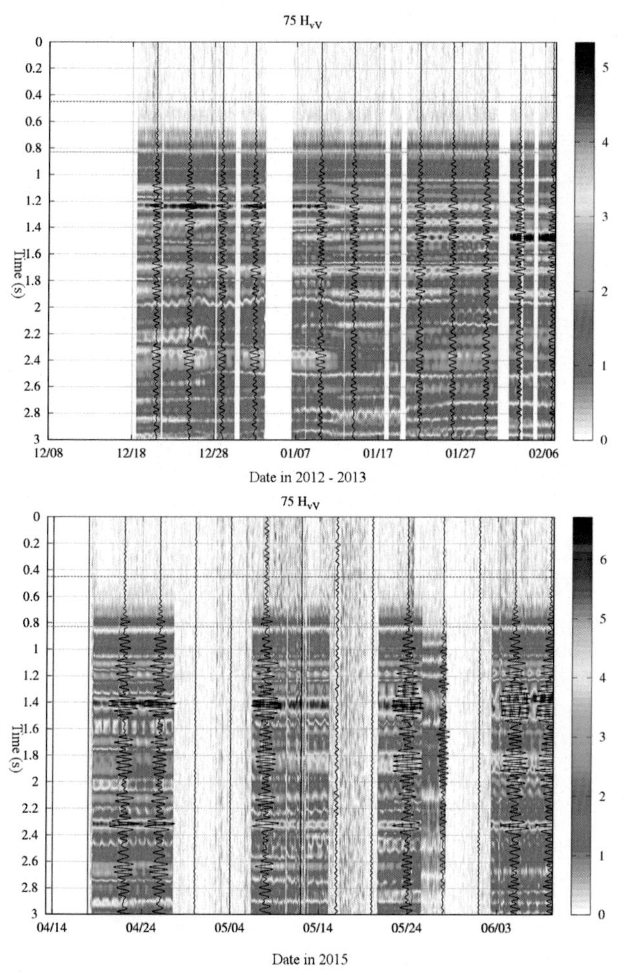

Figure 7.31 Results of H_{vV} at station #75 in December 8, 2012—February 6, 2013 (top), April 14—June 10, 2015 (bottom). *Red* and *green* horizontal lines show expected travel times of 3.5 km/s and 1.9 km/s, respectively. The offset distance of the station is 1.579 km.

interstation transfer function is obtained from the cross-power spectra of the noise frequencies averaged for a certain period. The inverse Fourier transform gives the time-domain waveform corresponding to the cross-correlation.

Kasahara et al. (2014c) applied this method to the Saudi Arabia ACROSS. Fig. 7.32 shows the results of seismic interferometry using stations #25 and #51. We can see very strong coherent arrivals

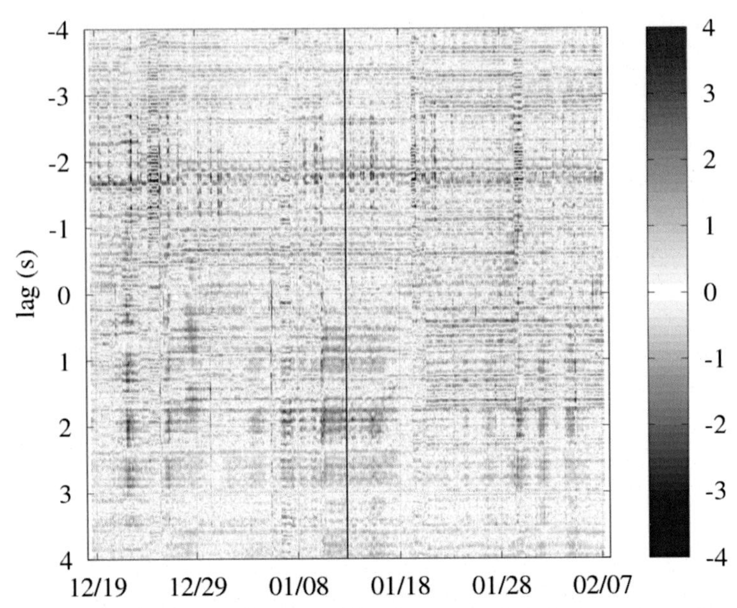

Figure 7.32 Time-lapse waveform resulting from seismic interferometry using background noises observed at #25 and #51 with distance of ~2500 m. Each trace was obtained by stacking for 48 h. The vertical axis is the lag time in seconds. A positive lag means southward propagation. The horizontal axis is the date from December 19, 2012, to February 7, 2013. The largest phase at −2.0 s was interpreted to be a Rayleigh wave.

at −2.0 s, which correspond to Rayleigh wave. A Rayleigh wave propagating northward was clearly identified for most of the station pairs. They considered the dominant noise source to be traffic noise from the highway south of the test site.

In this section two time-lapse observations were obtained: one with an active seismic source (ACROSS) and another with background noise as a passive seismic source. The results obtained from the active seismic source were more complicated than the results from the passive source.

The surface waves observed from the active and passive sources had different frequency contents, so we could observe phenomena at different depths. The temporal changes in the later phases observed with the active source showed a good correlation with the air temperature obtained by each data logger, which implies changes in a very shallow part of the ground. We identified similar waveform changes due to near-surface temperature changes with our experiment in Japan (Kasahara et al., 2014b).

7.4.7 Summary of Time-Lapse Study in Saudi Arabia

The time-lapse experiment in Saudi Arabia from December 2012 to February 2013, and April to December 2015, were carried out using an accurately controlled continuous seismic source. Very distinct temporal changes were identified in transfer functions for the vertical and horizontal forces (vibrations) at all 31 stations. Because there was no change in the source spectra during this period (see Chapter 9), the main cause of the waveform changes to the transfer (Green's) functions may be changes in the water table during this period. There are 64 pumping wells at the Al Wasse field, and they may have drastically changed the water table distribution.

CHAPTER 8

Near-Surface Effects

Contents

Near-surface effect is one of the big issues on the time-lapse study on shore. In the case of offshore, it is unknown whether the near-bottom sediments affect the time-lapse problem or not. Onshore, the factors that affect the seismic characteristics at near surface could be temperature change, seasonal variation, and precipitation variation caused by rainfall.

8.1 EFFECT OF PRECIPITATION

We explain the effects of precipitation by referring to cases in Japan. A test using Accurately Controlled and Routinely Operated Signal System (ACROSS) seismic source was done in Awaji Island, Japan. The Great Hanshin earthquakes (Mw = 6.9) occurred in 1995 along the fault between Kobe city and Awaji Island. The main shock initiated between Awaji Island and Kobe city and propagated to Kobe city where human lives and construction greatly suffered. The Nojima earthquake fault appeared on the ground surface in Awaji Island. To investigate the fluid flow along the fault, 375 t water (25 t/day) was injected to 550 m depth in March and May 2003 (Fig. 8.1). To study the effect of water injection to the fault, time-lapse measurements using an ACROSS and a small seismic array were carried out by researchers of Nagoya University. They reported that there were not distinct changes associated to the water injection although reanalysis using the original dataset revealed focusing of waveform residuals near the injection area (Kasahara et al., 2012). In the ACROSS data analysis, Misu et al. (2004) found good correlation

Time Lapse Approach to Monitoring Oil, Gas, and CO_2 Storage by Seismic Methods
ISBN 978-0-12-803588-7
http://dx.doi.org/10.1016/B978-0-12-803588-7.00008-X
Copyright © 2017
Elsevier Inc.
All rights reserved.

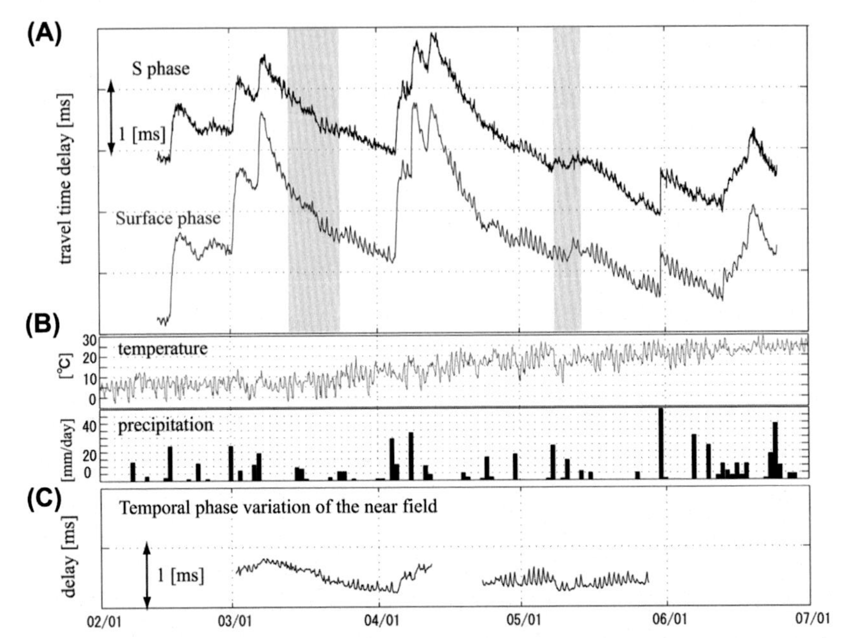

Figure 8.1 (A) Delays of S- and surface wave phases. *Gray* shows water injection nearby borehole at 550 m depth. (B) Temperature variation and precipitation measured by Japan Meteorological Agency in Awaji Island. (C) Temporal variation measured near the source. *After Misu, H., Ikuta, R., Yamaoka, K., 2004. Active monitoring of upper crust using ACROSS-seismic array system, Abstract of Annual Meeting of Japan Geoscience Union.*

between S- and surface wave travel-time changes and precipitations. When the precipitation by rainfall was greater than 20 mm/day, the S- and the surface waves showed ~2 ms delay and the effect continued for a half month.

Fig. 8.2 shows the change of water flow speed at different depth by rainfall (Kasahara et al., 2011d, 2012). The permeability of ground layer was assumed. If the precipitation given by Misu et al. (2004) is assumed, the patterns of flow rate show different from shallow to deep. The pattern of flow rate at 5 m depth is similar to the travel-time delays seen in S- and surface waves. The flow rates at deeper than 5 m show smoother patterns. However, the effect of precipitation continues approximately 15 days, which is the same duration as the observation by Misu et al. (2004). The 375 t water injection to 550 m depth in 2003 seems not so huge compared to the rainfall. If rainfall in the area of 1 × 1 km and the precipitation is 1 cm, there is

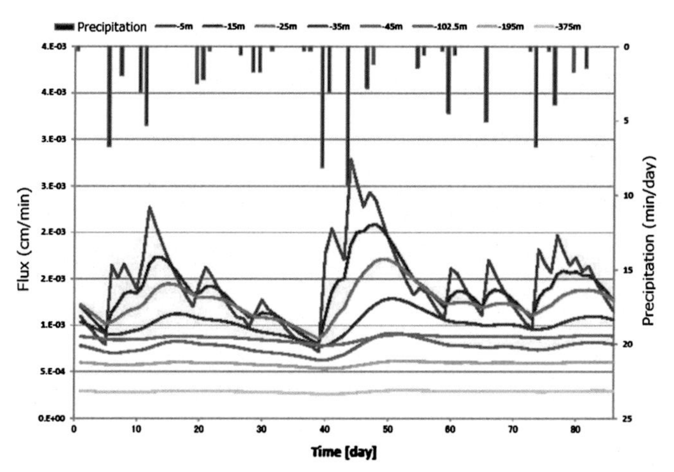

Figure 8.2 Comparison of precipitation and S-travel-time change in Misu et al. (2004) (Kasahara et al., 2011d, 2012).

10,000 t water on the ground. This suggests the precipitation could be one of the large factors affecting the time lapse.

Another example was also obtained in Awaji Island by the study in 2011 (Kasahara et al., 2011d) (Fig. 8.3). The major result of the 2011 study is given in Section 7.2. The temporal variation of the first arrivals of P was obtained by cross-correlation between the waveform in February 21 and the current ones (Figs. 8.3 and 8.4). The stations #17 and #18 are 50 m apart and showed similar rainfall effects on amplitudes of P first arrivals. The precipitation in this area was 3 mm on February 28 and 2 mm on March 7. Even though the precipitation was not high, the amplitude changes were clearly seen at both stations. On the other hand, the travel-time changes larger than 0.2 ms were not identified on February 28, but a noticeable change with approximately 0.2 ms travel-time delay was identified at both stations. Apart from that, the important thing to note is the duration of precipitation effects. Even though the precipitation was only 5 mm/h, the effect appears on amplitude and continued for several days. In these examples, travel-time changes of P are identified but S part is difficult to separate from the P coda parts.

Meersman (2013) studied the near-surface time-lapse effect in Peace River, Alberta. He separated upgoing and downgoing S-waves using vertical and radial components and found that upgoing S-wave

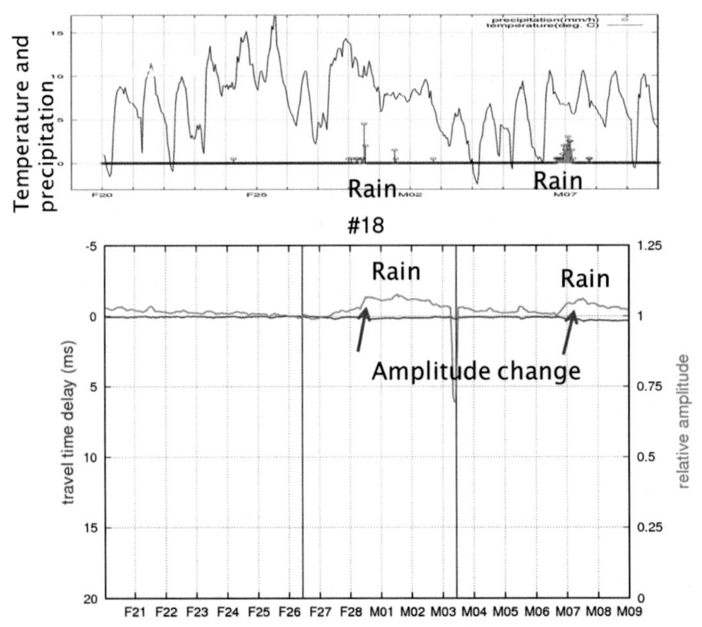

Figure 8.3 (Top): thin curve is temperature (°C) and thick line is precipitation (mm/h). (Bottom): the travel-time changes (ms) and amplitude changes of first-arrival P at the station #18 in Awaji study (see Section 7.2 for details). Horizontal axis is time from February 21 to March 10, 2011.

time-lapse did not change, but downgoing time lapse showed changes. He thought the change was caused by gradual drying of upper 12-m-thick soil.

8.2 EFFECT OF TEMPERATURE

There is no distinct effect on temperature variation except at the frozen temperature. In Figs. 8.3 and 8.4, you cannot see large effects of temperature change between 0°C and 15°C. However, we can see small daily variation in the records of #17 (Fig. 8.4). We also noticed the daily variation on observed records of the green tuff region (Fig. 7.20) and Al Wasse (Figs. 7.29 and 7.32). In the later chapter of repeatability discussion, the normalized root mean square of the ACROSS seismic source showed daily variation of the geophone in the source room. The cause of daily variation could be temperature change during daytime and nighttime. The geophone in the seismic source room was placed at

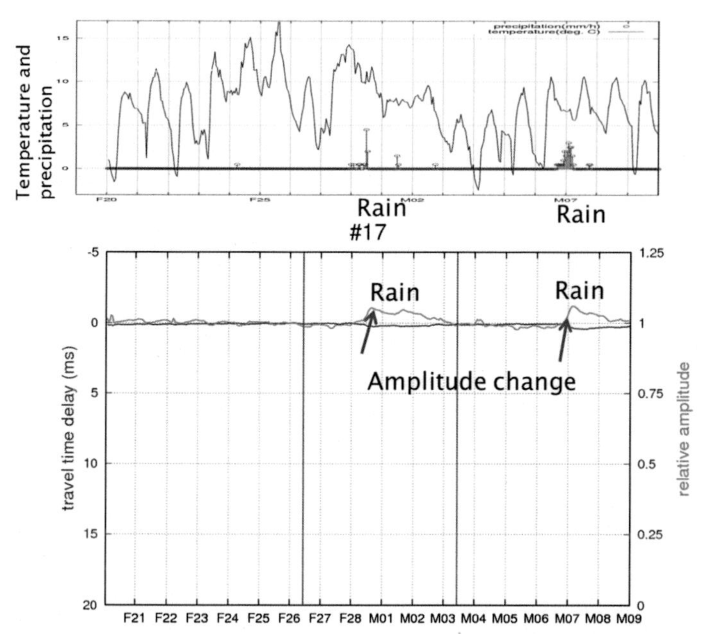

Figure 8.4 (Top): thin curve is temperature (°C) and thick line is precipitation (mm/h). (Bottom): the travel-time changes (ms) and amplitude changes of first-arrival P at the station #17 in Awaji study (see Section 7.2 for details). Horizontal axis is time from February 21 to March 10, 2011.

the basement of the source and the room temperature was kept constant. Considering the circumstance of the geophone location, the elastic response of the source room itself could be changed.

Other temperature effects are freezing of near-surface layers. If the temperature comes to below the freezing point, the freezing of near-surface layer could happen. Fig. 8.5 shows the waveforms during the winter in green tuff region in Japan. It is clearly seen that the waveforms around 0.4 and 0.7 s change. The phase at 0.4 s corresponds to S-phase. When the air temperature was below 0°C at midnight, amplitudes around 0.4 s increased. This can be explained by freezing of near-surface layer. On the other hand, the P first arrivals did not show noticeable change. Fig. 8.6 is the record obtained by borehole. Although the depth of geophone is 200 m, the later arrivals than P phase showed the waveform changes associated with the temperature variation crossing 0°C at the travel times of 0.6 s and 0.82 s. Because the geophone was located at

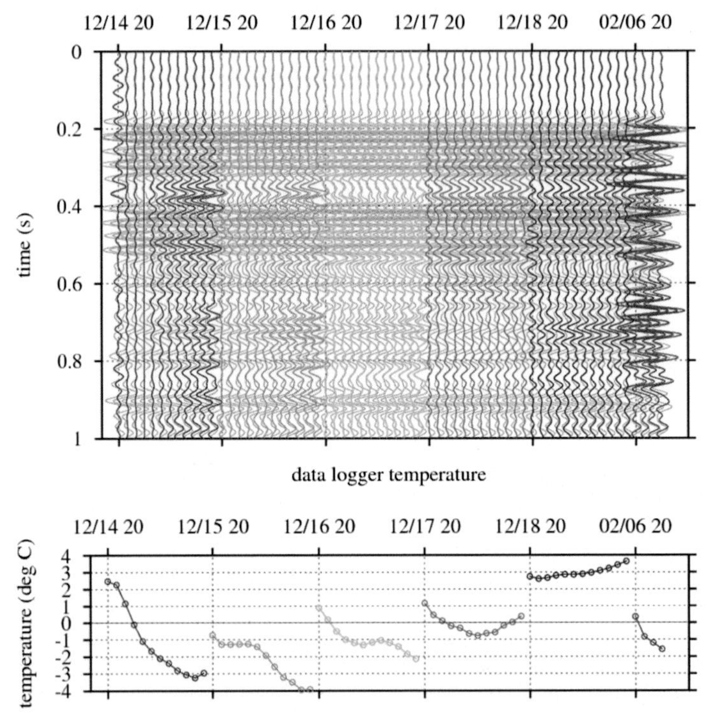

Figure 8.5 (Top) Transfer functions of every hour at 378-m distance from the seismic source from December 14, 2013, to December 18 and February 6, 2014. Gray wiggles are the original waveforms and colored ones are differences from the trace of December 14, 2013, at 23:00. Vertical axis is travel time and horizontal axis is time. Every 12 h operation of the source from 8 p.m. to 8 a.m. (Bottom) temperature measured at nearby weather station. Vertical geophone was buried at 30 m in the ground.

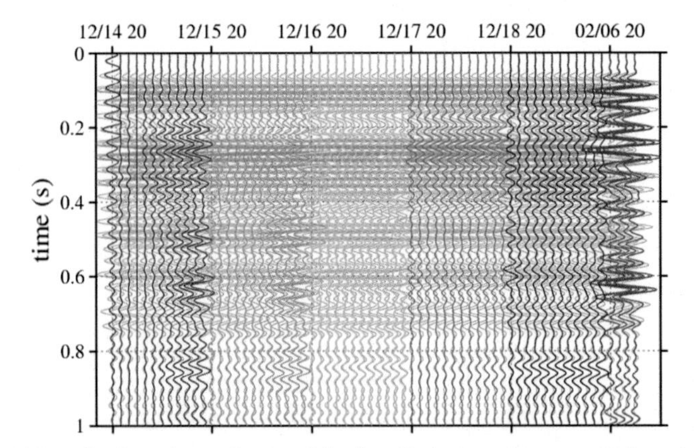

Figure 8.6 Transfer functions observed by borehole geophone at 200 m depth. *Gray* wiggles are the original waveforms and colored ones are differences from the trace of December 14, 2013, at 23:00.

200-m borehole, direct effect of near-surface layer could not be the cause. The cause could be that tube-wave transmission was changed by freezing effects.

8.3 GROUND ROLLS (SURFACE WAVE) EFFECTS

The ground rolls that are surface waves such as Rayleigh and Love waves could mask the target reservoirs. If the seismic source is vertical vibrator, the reflections from the deep-seated reflector could be dominant in vertical component. On the other hand, the particle motion of Rayleigh wave is rotational. Ground rolls have characteristics of very large amplitudes elliptical plane polarization, strong

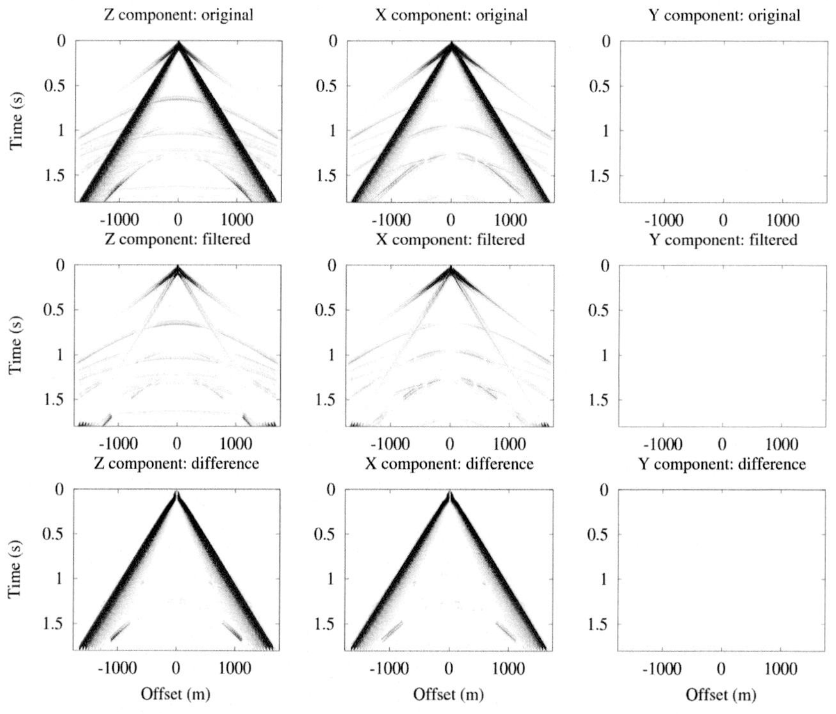

Figure 8.7 Application of Chen et al. (2013) algorithm to remove Rayleigh waves for synthetic waveforms. (Top): z and x (radial) components of synthetic waveforms. (Middle): applied the polarization filter of Chen et al. (2013) algorithm. (Bottom): residuals of original one and after the application of polarization filter. The vertical axis of each diagram is travel time in seconds and the horizontal axis is offset distance from the source.

energy, low apparent velocity, and low frequency. Using these characteristics, we could remove or suppress.

Chen et al. (2013) proposed the method of polarization analysis using the complex singular value decomposition (CSVD) technique to suppress the surface wave. Polarization analysis uses the characteristics of Rayleigh wave where the particle motion is orbital. During the orbital motion, vertical component and radial component could be represented by complex time-variant one principal component with orthogonal phases. Chen et al. (2013) used to extract orthogonal three time series using CSVD algorithm.

We tested this method to synthetic data shown in Fig. 8.7. By this method, surface waves seen in vertical and horizontal components were almost removed. The residual waveforms in the bottom figure

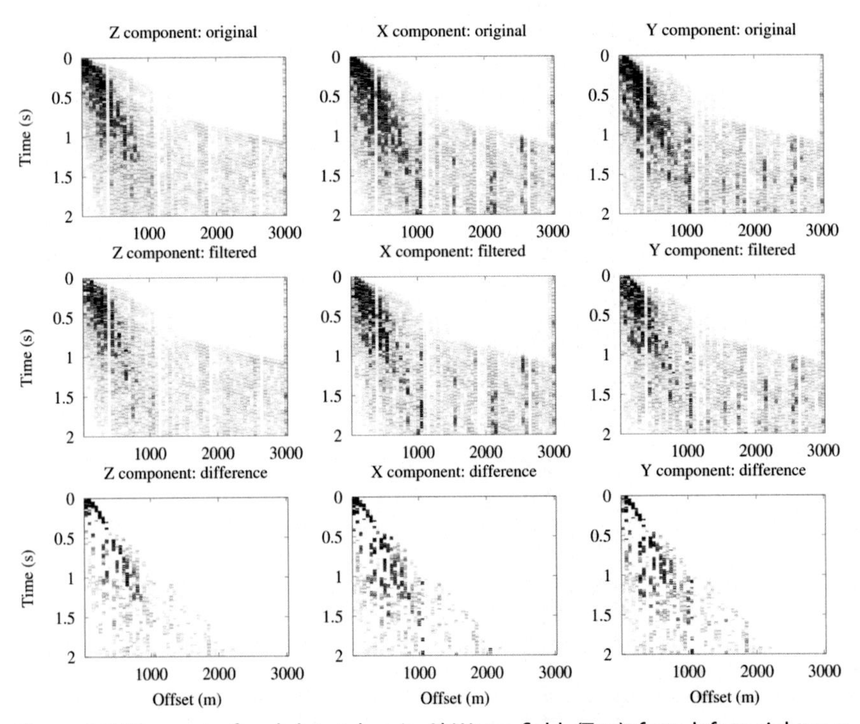

Figure 8.8 The case of real data taken in Al Wasse field. (Top): from left to right; z, x (radial), and y components of waveforms. (Middle): applied the polarization filter of Chen et al. (2013) algorithm. (Bottom): residuals of original one and after the application of polarization filter. The vertical axis of each diagram is travel time in seconds, and the horizontal axis is offset distance from the source.

show large amplitudes of Rayleigh waves. Using the real data taken in Al Wasse field, we tested this method as shown in Fig. 8.8. In the real-data application, most surface waves could be removed, but a part of other phases were removed.

This method seems useful to separate body waves and the Rayleigh wave, but three-component seismic data are required.

CHAPTER 9

Repeatability

Contents

9.1 FACTORS CONTROLLING REPEATABILITY

The repeatability in the time-lapse study is the most important factor to control the accuracy of measurements. Meunier (2011) suggested that factors such as equipment, coupling, positioning, propagating medium, and ambient noise affect the repeatability. As described by Johnston (2013), small errors in lateral positioning and in time shifts rarely impact the 3D seismic imaging, but they can influence repeatability and the interpretation of the time lapse.

The repeatability is controlled by several factors as follows (also see Table 1 in Section 1.4).

1. Source signature changes
2. Ground coupling
3. Media through paths
4. Geophones
5. Overburden differences due to errors in positioning
6. Time base accuracy
7. Ambient noise

Time Lapse Approach to Monitoring Oil, Gas, and CO_2 Storage by Seismic Methods
ISBN 978-0-12-803588-7
http://dx.doi.org/10.1016/B978-0-12-803588-7.00009-1
Copyright © 2017
Elsevier Inc.
All rights reserved.

9.2 NORMALIZED ROOT MEAN SQUARE AND PREDICTABILITY

Kragh and Chirstie (2002) proposed normalized root mean square (NRMS) for the repeatability,

$$\mathrm{NRMS} = \frac{200 \times \mathrm{RMS}(a_t - b_t)}{\mathrm{RMS}(a_t) + \mathrm{RMS}(b_t)} \tag{9.1}$$

where the RMS operator is defined as:

$$\mathrm{RMS}(x_i) = \sqrt{\frac{\sum_{t_1}^{t_2} (x_t)^2}{N}}. \tag{9.2}$$

N is the number of samples in the interval $t_1 - t_2$.

If both traces contain random noise, the NRMS value is 141%. If both traces anticorrelate, the NRMS error is 200%. If one trace is half amplitude of the other, the NRMS error is 66.7%.

They showed the NRMS depends on location error of geophones.

Another definition for the repeatability proposed by Kristiansen et al. (2000) is coherence in White (1980).

The predictability (PRED) is defined as:

$$\mathrm{PRED} = \frac{\sum \Phi_{ab}(\tau) \times \Phi_{ab}(\tau)}{\sum \Phi_{aa}(\tau) \times \Phi_{bb}(\tau)} \tag{9.3}$$

where Φ_{ab} is cross-correlations between traces Φ_a and Φ_b within time window t_1 and t_2. The coherence of two traces is obtained by summed squared cross-correlation within a time window divided by the summed product of the two autocorrelations.

The NRMS and PRED is in negative correlation (Kragh and Chirstie, 2002).

In the case of the Gulf of Mexico, the best NRMS values were 18–30% and most of the errors were positioning errors (Kragh and Chirstie, 2002).

Landrø (1999) showed the rms repeatability is from 20% to 100% with variation of separate distance between two shots. He thinks that the higher errors are associated with highly heterogeneous area. In most of 4D seismic survey 10–30% is thought to be typical good value (Johnston, 2013). Eiken et al. (2003)

obtained approximately 40% NRMS by two surveys with 25-m lateral offset.

Jervis et al. (2012) obtained 15–20% of poststack repeatability using 30-m buried geophones.

Cantillo (2011) criticized the use of NRMS and PRED as repeatability metrics. He thinks two values are simply a combination of distortion and time shift and proposed to use signal to distortion (SDR), which quantifies energy in the 4D difference after the time shifts have been removed. Eiken (2003) theoretically showed that repeatability decreases with increasing frequency. Johnston (2013) thinks that PRED is sensitive to the length of correlation window. He thinks that the comparison of absolute value of repeatability measures among 4D projects is not particularly meaningful.

Meunier (2011) claimed the permanent receivers (and some permanent sources) are much more stable than redeployable receivers or moving source. Schissele et al. (2009) used the piezoelectric vibrator recorded by buried geophones and obtained very good repeatability.

The factors relating to the source signature could be minimized by good source control similar to Accurately Controlled and Routinely Operated Signal System (ACROSS). The method to keep the best repeatability on ACROSS source is described in Chapter 3.

If the source position is fixed as in the case of ACROSS, the positioning error of source is minimized. The coupling of the ground and the source is a serious problem in vibrators (e.g., Vibroseis). The force to the ground can be measured by accelerometer to feedback to the control system in most of vibrators. However, the coupling condition is not easy to make corrections. Although the cross-correlation of the source signature measured by nearby sensor(s) and the receiver might reduce the error on repeatability, it still brings some inaccuracy on the repeatability.

9.3 SOURCE SIGNATURE REPEATABILITY

If a seismic source on shore is a kind of sweep-type vibrator (e.g., Vibroseis), each sweep could be controlled by feedback system based on a design sweep pattern and frequency base referring to very accurately controlled time base. The force (acceleration) could be calculated by measurement of accelerometers on baseplate and loading mass. The inaccuracy on oil pump could cause some error. Although these factors can be minimized by measuring acceleration on the baseplate, the coupling of baseplate to the ground could cause large errors on source signature. The cross-correlation of nearby geophone records or baseplate acceleration might improve the repeatability errors.

Any impact source, explosives, or land air gun could have larger repeatability errors.

For the air guns offshore, the seismic source signatures can be estimated by the synthetic waveforms based on air gun pressure, delay response of electromagnetic switch, water depth of air guns, towing method, and/or tuning way of gun array. The water depth, sound speed due to water temperature and salinity, and tidal height can be estimated with careful treatment by theoretical or empirical relations.

If a seismic source on shore is buried in heavy base similar to the case of ACROSS, the error due to the coupling could be minimized.

9.4 GROUND COUPLING

The situations onshore and offshore are different.

Onshore the coupling of source to ground is one of large factors to control the repeatability. If source moves 5 m, the overburden changes might occur. Even if the source is almost in the same location, the small change of baseplate of vibrator might cause different source signature. When monitoring by accelerometer on the baseplate it is difficult to compensate for the change of source signature.

9.5 STRUCTURE BETWEEN SOURCE(S) AND RECEIVERS

The water contents in the near-surface layer might greatly affect the repeatability. The precipitation and snowfalls might control the water in near-surface layer. Misu et al. (2004) found a similar pattern between rainfalls and travel-time change of arrivals. Kasahara et al.

(2011) suggested the water flow at the layer shallower than 5 m may determine the travel-time change of S-wave and surface wave. If the atmospheric temperature is below 0°C, the water in the surface layer will be frozen and P and S velocity can be changed. Kasahara et al. (2014) found the arrival-time changes when atmospheric temperatures are close to 0°C. Saiga et al. (2006) found similar results in winter using ACROSS source at Tono, Japan.

9.6 GEOPHONES

If plural time-lapse measurements are not exactly the same equipment, it is necessary to apply the corrections for seismic sources and receivers. It is necessary to know the response characteristics of each geophone for correction.

The repeatability of receivers could depend on response of geophones and coupling to the ground. The near-surface layer just beneath the geophones could be affected by water contents. The multiple reflections in the surface layer and/or surface waves largely affect the receiving waveforms. To avoid the temporal change due to velocity change in the near surface, burying geophones down to several tens of meters could be effective. Secular change of geophones should be corrected, but the calibration of each geophone response may be difficult.

Even though geophones are installed at the borehole, later arrivals of P and S first breaks may change by reflections bounced at the ground surface (Kasahara et al., 2014).

9.7 POSITIONS OF SOURCE(S) AND REVISER(S)

In marine 4D seismic survey, the most important factor in repeatability is acquisition geometry (Calvert, 2005). Differences in source and receiver positions cause different paths. The earth response of media between source(s) and receiver(s) could include large errors due to near-surface effect. Eiken (2003) obtained significant difference of 25-m laterally apart seismic reflection profiles. He obtained ~40% NRMS for the two traces and even the lateral shift is 10 m, the NRMS is as high as 20%. Considering the survey conditions onshore and offshore, 20–40% NRMS might be evitable. Landrø (1999) obtained 8% of NRMS for two shots and the same receivers with 5-m-distance separation at the VSP experiment. He estimated NRMS for 87, 125 shots pairs and obtained NRMS distributing from 20% to 100%. Higher errors might be associated with highly heterogeneous overburden (Misaghi et al., 2007).

To avoid the overburden difference caused by positioning errors onshore, fixed source and receiver position are desirable. In ACROSS seismic source, the source is mounted in the heavy concrete block (~100 t) at fixed position(s).

9.8 TIME BASE AND DIGITIZING RESOLUTION

The time is one of control factors for the time-lapse study. If the very accurate time standard such as GPS clock is used, it will be minimized. The sampling rate for the data acquisition system should be very accurate and the fast sampling as 1 ms is required.

9.9 AMBIENT NOISE

Pevzner et al. (2011) examined the repeatability of CO_2 CRC Otway data. They examined the weight reflection surveys using 1320 kg weight drop in June 2007 (wd07) and November 2008 (wd08) and IVI Minivib in November 2008 (mb08). The repeatability using the combination of wd08/wd07, wb08/mb08, and mb08/wd07 was more than 20%. The weather conditions in June 2007 and November 2008 were wet and dry, respectively. They also calculated the signal-to-noise ratio (S/N) for the same dataset and found that the greater repeatability corresponds to those with higher S/N. There is the linear relationship between the S/N and NRMS, and they think the ambient noise ruled their NRMS.

9.10 REPEATABILITY OF ACROSS SOURCE

One ACROSS seismic source station and 32 geophone stations in Al Wasse field were used for the time-lapse and repeatability studies (Kasahara et al., 2016a,b,c). The ACROSS source in Al Wasse was

Figure 9.1 Surface layer near the ACROSS source and geophones.

located in the Saudi National water pumping station. The geology of the site comprises limestone, sandstone, and unconsolidated sands. The surface layer in the Al Wasse site is shown in Fig. 9.1. They used the frequency sweeps of ACROSS from 10 to 50 Hz in the first period (December 2012 to February 2013) and 10–40 Hz in the second period (April–June, 2015) within a 200-s time window.

To obtain the travel-time variation (dT) and the amplitude variation (dA) of geophones exited by the ACROSS, the cross-correlation of *P* first arrivals was used.

The source and receiver map are shown in Fig. 9.2. There are more than 64 water pumping wells to pump up water continuously from aquifers located around 400 m depth. It seems that pumping up the water from the aquifers strongly influences the dT, dA, and NRMS variations. It might cause degradation of the apparent repeatability.

The NRMS repeatability based on Eq. (9.1) was calculated using a geophone in the ACROSS source room (Fig. 9.3). The NRMSs during 12 days were less than 2%. We can clearly identify daily variations from Fig. 9.3. Because the room temperature was kept constant, the cause of daily variation could be housing deformation due to the large temperature variation of outside air. The ACROSS seismic source was in the desert area and the outside temperature was very low during nighttime in winter.

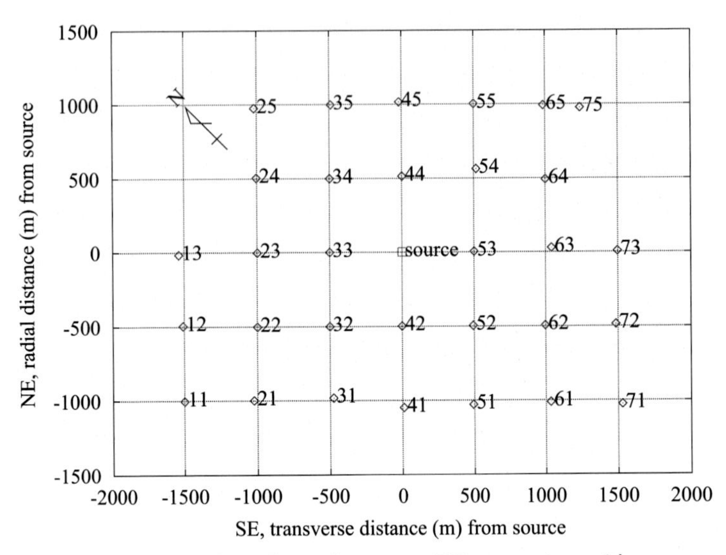

Figure 9.2 Map of geophones on 500-m spacing grids.

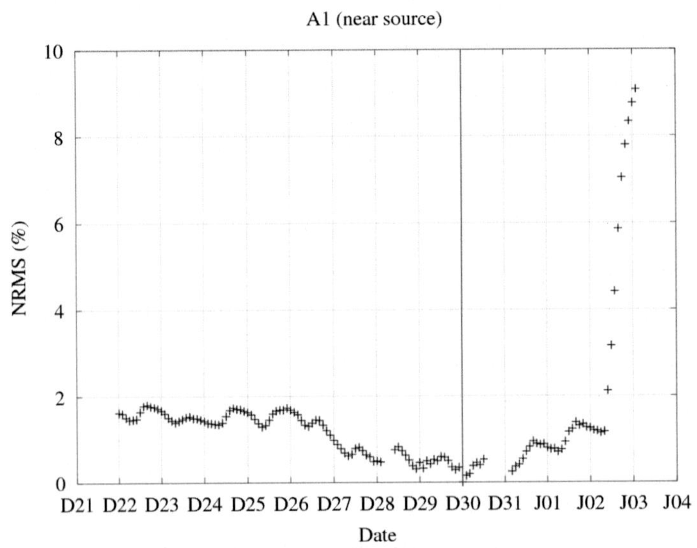

Figure 9.3 NRMS using geophones in the ACROSS source room during December 22, 2012, to January 2, 2013. Reference date is December 30, 2012. Clear daily variation was observed. The NRMS is less than 2% during these periods.

Next, dT and dA were calculated. The cross-correlation of 50-ms time window around P-wave first arrival was used to calculate the dT and dA. The dT and the dA in #33, #53, #44, and #52 stations show small temporal variation during two separate periods. There are two years between the first test and the second test. The dT of #33 for the first and second periods was less than ± 0.2 ms, but it was a bit greater in the second period as ~1ms (Fig. 9.4). The dA of #33 in the first and the second periods was within a few percent for the first and second periods. The waveform variation is shown in Fig. 9.5. The dT variation of #53 in the first period was less than ± 0.2 ms (Fig. 9.6). The dA variation of #53 was within a few percent in the first period and less than 10% in the second period. The dT and dA variations and waveform variations of #44 are shown in Figs. 9.7 and 9.8, respectively. There was a shift of dT and dA from the first period and the second period (Fig. 9.7). The dT shift was not so large, but dA in the second period decreased 20% compared to one in the first period. The dT variation of #52 in the first period is ca. ±0.4 ms, but the dT in the second period showed +1 ms shift to the first period

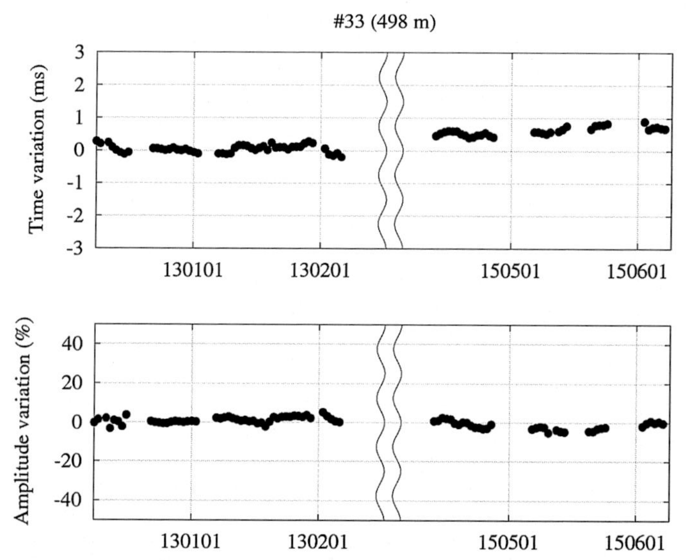

Figure 9.4 Daily variation of dT and dA of *P* phase using cross-correlation of #33 station (distance is 498 m). The dT for the first and second periods was less than ±0.2 ms, but it was a bit greater in the second period as ~1 ms. The dA in the first and the second periods was within a few percent for the first and second periods.

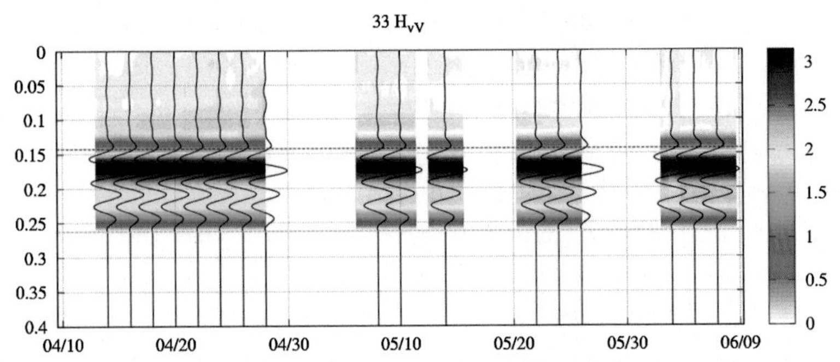

Figure 9.5 An example of *P* waveforms variation of #33 station in April 12–June 09, 2015, traveling with 3.4 km/s. Vertical force by ACROSS (10–40 Hz) seismic source. Geophone: 1-Hz vertical component. The records are transfer functions between source and geophone at surface of ground.

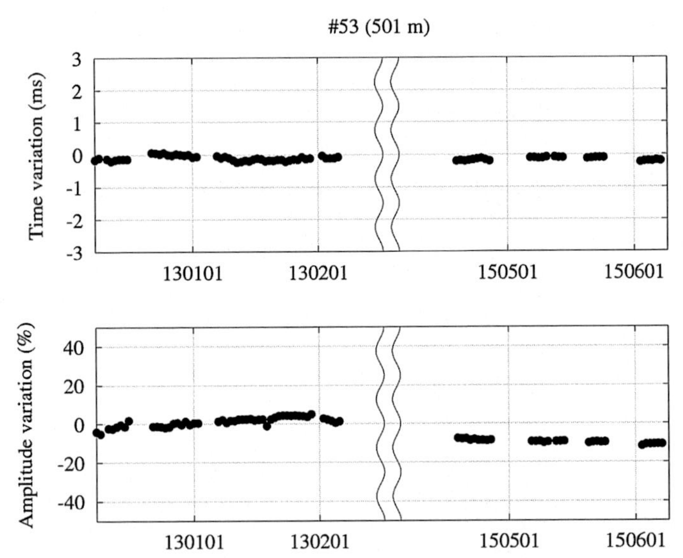

Figure 9.6 Daily variation of dT and dA of *P* phase using cross-correlation at #53 station (distance is 501 m). The dT variation in the first period was less than ±0.2 ms. The dA variation was within a few percent in the first period and less than 10% in the second period.

(Fig. 9.9). The dA showed similar variation to dT. It seems some change in #54 occurred in the subsurface because such dT and dA in #52 shifts were not seen in #33 and #53 stations.

The dT and the dA of eight stations within 750 m distance from the ACROSS source are summarized in Fig. 9.10. Some stations

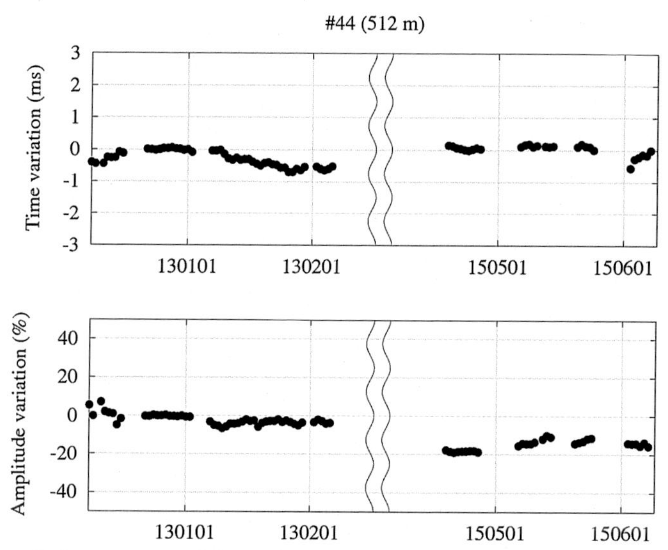

Figure 9.7 Daily variation of dT and dA of *P* phase using cross-correlation at #44 station (distance is 512 m). The dT variation in the first period was less than ±0.5 ms. The dT in the second period showed +0.5 ms shift. The dA variation was within a few percent in the first period, but dA decrease in the second period.

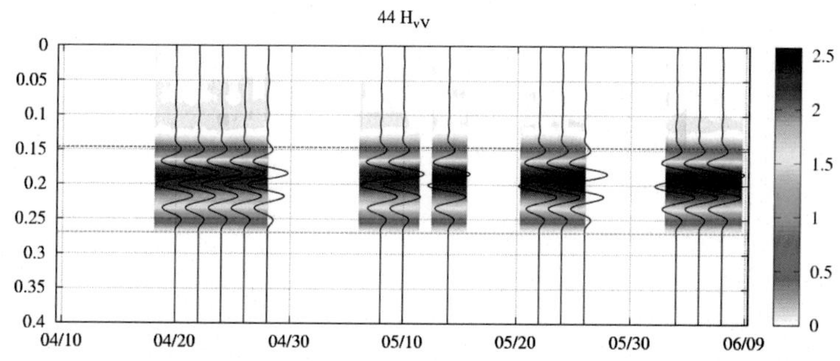

Figure 9.8 Temporal variation of waveforms in April 18 to June 09, 2015 observed at the #44 station. Offset distance is 512 m. It is difficult to see any temporal changes on the waveforms.

show very large temporal variation. The reason could be amplitude of P arrivals and continuous water pumping. The P arrivals disappear at the distance further than 700 m due to the presence of the low-velocity layer at the depth of 100 m in Al Wasse region. The layer between 100 m and 900 m is sandy with aquifer. The water in the aquifer could move in this thickness by water pumping works.

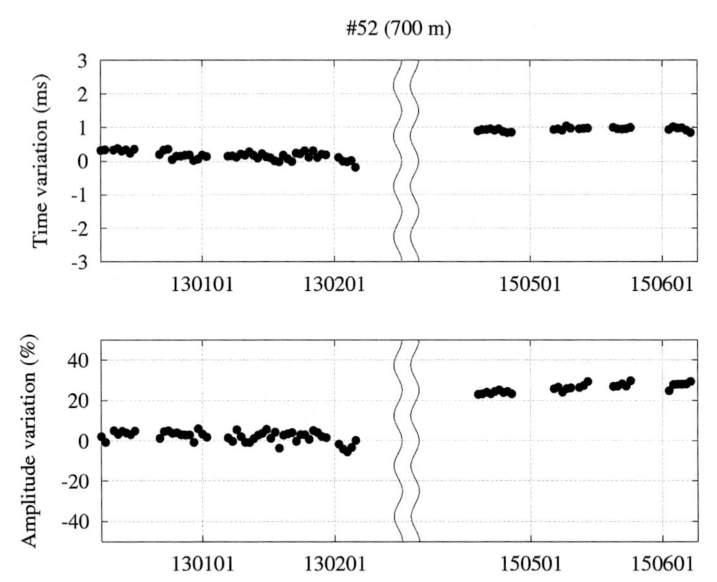

Figure 9.9 Daily variation of dT and dA of *P* phase using cross-correlation of #52 station (distance is 700 m). The dT variation in the first period is ca. ±0.4 ms, but the dT in the second period showed +1 ms shift to the first period. The dA showed similar to dT effect. It seems some change occurred in the subsurface because such dT and dA shifts were not seen in #33 and #53 stations.

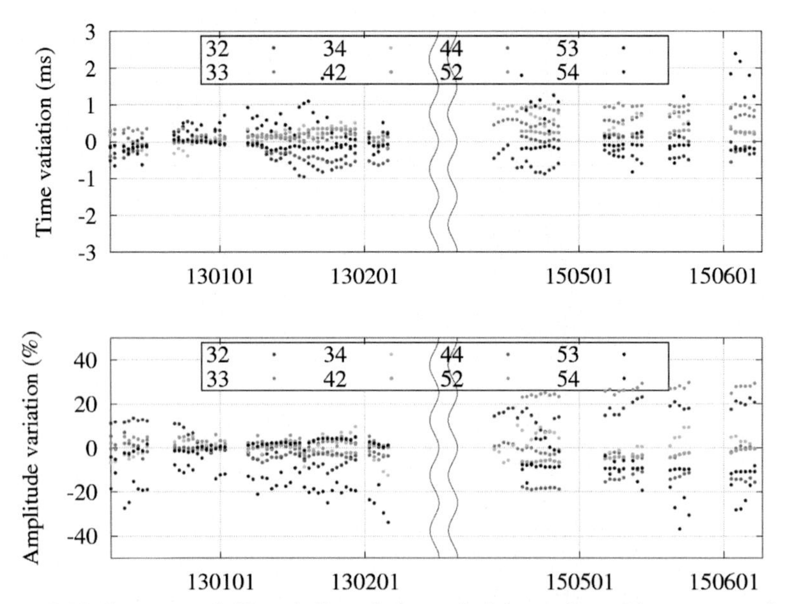

Figure 9.10 Summary of dT and dA variations of eight stations. Distances are from 500 m to 750 m. There are places of extremely small temporal changes and large changes due to weakness of P arrivals and subsurface changes due to water flow in the aquifer.

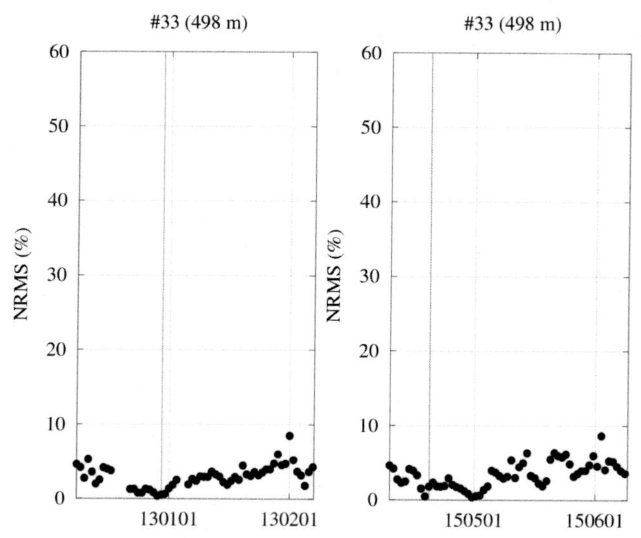

Figure 9.11 NRMS variation of #33. Reference dates are December 30, 2012, and April 19, 2015. The NRMS is approximately less than 5% during one to two months.

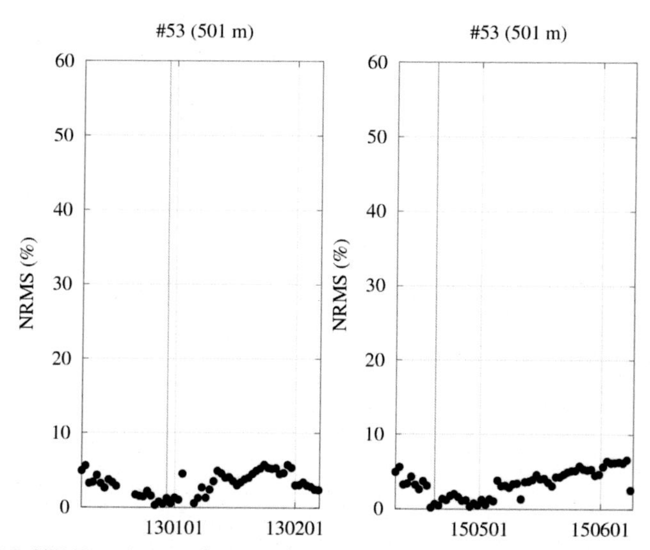

Figure 9.12 NRMS variation of #53. Reference dates are December 30, 2012 and April 19, 2015. The NRMS is slightly greater than 5% during one to two months.

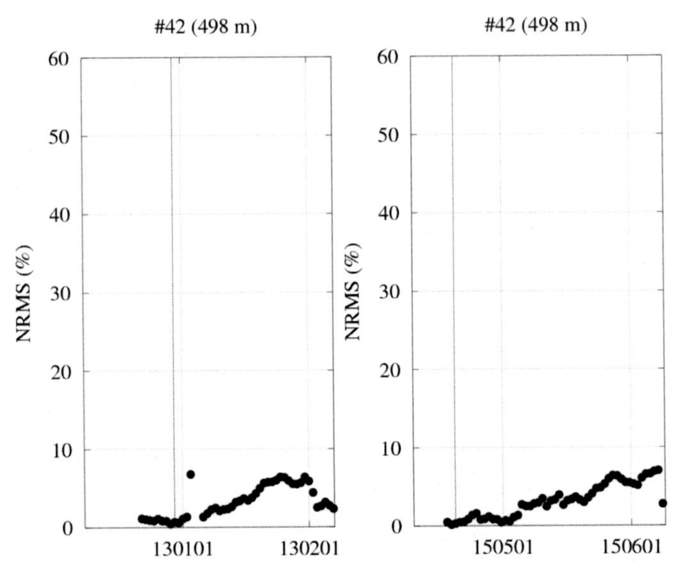

Figure 9.13 NRMS variation of #42. Reference dates are December 30, 2012 and April 19, 2015. The NRMS is approximately within 7% during one to two months.

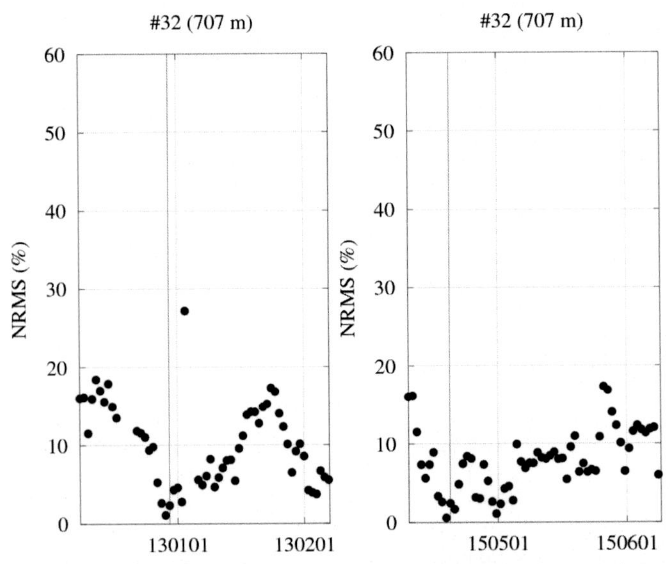

Figure 9.14 NRMS variation of #32. Reference dates are December 30, 2012 and April 19, 2015. The NRMS is approximately within 10–20% during one to two months, but the variation was continuous.

Next, NRMS, Eq. (9.1), was calculated in the following way (Figs. 9.11–9.14): For the stations within 1 km from the source, the transfer function using 2-h data are stacked with weights for sliding one-day window. The P-wave arrival-time windows with the width of 50 ms assuming apparent velocity of 3.5 km/s are applied. The NRMSs are calculated with the reference traces. The one-day medians of NRMSs are calculated.

Because of the different source signatures in 2012–2013 and 2015, the NRMS was calculated using different reference date. The sweep in the first period was 10–50 Hz and the one in the second period was 10–40 Hz.

NRMS variations of stations #33 (Fig. 9.11) and #53 (Fig. 9.12) were smaller than 5% in the fist and the second periods, which is a similar result seen in dT and dA. Considering the NRMS of records in the source room, the temporal variation of source signal itself is small (~less than 2%) and the environmental change such as pumping, temperature change affect to the NRMS of grid stations.

NRMSs of other stations show very large variation suggesting that this is not due to the source signature change.

Because the geophones were located at the surface and the test field has more than 64 water pumping stations, the apparent NRMS observed by the grid stations showed great temporal variation. However, the temporal variations are time lapse itself.

In summary, the ACROSS test field is located at the water pumping stations with more than 64 wells. The depth of the aquifer is ~400 m and it affects the seismic records. The $V_P = 3.5$ km/s arrival quickly decays with distance and it disappears around 700 m due to the presence of low-velocity layer. The presence of aquifer and weak P arrivals could affect the NRMS estimation. The NRMS by the source room geophone showed less than 2% NRMS during 12 days. The NRMS of two stations were less than 5% during one to two months. The dT and dA of these two stations are extremely small, such as 0.2 ms within each term during two years. NRMS and dT and dA of other stations were large suggesting the effect to weak P arrivals and water pumping effects.

CHAPTER 10

Rock Physics

Contents

In the time-lapse study, the estimation of physical property change is important. The physical properties depend on materials, amount of porosity, temperature, and pressure. We will briefly review these effects. The V_P and V_S of rocks were measured under pressure by Birch (1960, 1961), Simmons (1964), Christensen (1965), and Hoshino et al. (2001).

10.1 PHYSICAL PROPERTIES OF POROUS MEDIA

The physical property change by mineral grain orientation and material contents in the porous media is estimated.

The estimation of isotropic elastic constants of crystalline aggregates was discussed by Voigt (1928), Reuss (1929), and Hill (1952). Voigt (1928) averaged the elastic stiffness over all lattice orientations using the assumption that strain is uniform throughout grains. Reuss (1929) averaged the elastic compliance assuming that stress is uniform throughout a grain. The Voigt method gives upper bounds of elastic moduli, and the Reuss method gives the lower bound of elastic moduli. The Hill method averages the upper and lower bounds (Sumino and Anderson, 1984).

The discussion on the isotropic elastic constants of anisotropic mineral aggregates can be applied to aggregates of different isotropic grains. Assuming uniform strain by the Voigt method or stress by the Reuss method, the following two averages are derived.

Time Lapse Approach to Monitoring Oil, Gas, and CO_2 Storage by Seismic Methods
ISBN 978-0-12-803588-7
http://dx.doi.org/10.1016/B978-0-12-803588-7.00010-8
Copyright © 2017 Elsevier Inc.
All rights reserved.

The Voigt averages $\overline{K_V}, \overline{\mu}_V$ are the upper bounds and are given by the following equations:

$$\overline{K_V} = \sum_i^N f_i K_i, \tag{10.1}$$

$$\overline{\mu_V} = \sum_i^N f_i \mu_i, \tag{10.2}$$

where K_i, μ_i, and f_i are bulk modulus, rigidity, and volumetric fraction for material i.

The Reuss average $\overline{K_R}, \overline{\mu_R}$ is the lower bound and is given by the following equations:

$$\overline{K_R^{-1}} = \sum_i^N f_i K_i^{-1}, \tag{10.3}$$

$$\overline{\mu_R^{-1}} = \sum_i^N f_i \mu_i^{-1}. \tag{10.4}$$

And Voigt-Ruess-Hill average is the arithmetic average of the Voigt and Reuss averages:

$$\overline{M_{VRH}} = (M_V + M_R)/2. \tag{10.5}$$

Instead of many different grains, we take the case of grains and fluid for porous media. The physical properties of porous media was discussed by Biot (1956a,b), O'Connell and Budiansky (1974), Berryman (1980), Mavko (1980), Gassmann (1951), and Takei (1998, 2002).

Gassmann's equation (1951) is

$$K = K_b + \frac{\left(1 - \frac{K_b}{k_s}\right)^2}{\frac{\phi}{k_L} + \frac{(1-\phi)}{k_s} - \frac{K_b}{k_s^2}}, \tag{10.6}$$

where,

K_b: volumetric compressibility of Skelton solid (Pa)

k_s: volumetric compressibility of solid (Pa)

k_L: volumetric compressibility of liquid (Pa)

ϕ: porosity

Gassmann's equation is the case of zero frequency limit for Biot (1956b).

According to Takei (2005), mean density of medium is

$$\bar{\rho} = (1-\phi)\rho_S + \phi\rho_L, \tag{10.7}$$

where,

ϕ: Volumetric fraction of liquid

ρ_S: Density of solid (kg/m^3)

ρ_L: Density of liquid (kg/m^3)

The V_P and V_S of aggregate are

$$V_P = \sqrt{\frac{K_b + \frac{4}{3}N + \frac{k_s(1-K_b/k_s)^2}{1-\phi-K_b/k_s+\phi k_s/k_L}}{\bar{\rho}}}, \tag{10.8}$$

$$V_S = \sqrt{\frac{N}{\bar{\rho}}}. \tag{10.9}$$

These equations are the same as Biot (1956a).

10.2 EFFECTS OF SHAPE OF PORE

The shape of the pores in porous media controls the velocity of aggregates (Takei, 2002). Berryman (1980) discussed the oblate spheroid pores. In this case the aspect ratio of pores changes the velocity. Mavko (1980) used tube-shaped pores, and O'Connell and Budiansky (1974) used the crack-shaped pores. Takei (1998) introduced the grain shape in porous media. Depending on the assumed shape of pores, the describing parameters are different. Takei (2002) proposed equilibrium geometry to crack. She summarized the different theories on the velocity estimation of porous media aggregates (Table 10.1).

Following Takei's description using the equilibrium geometry,

$$\frac{K_b}{k_s}(\phi, x) = 1 - \phi\Lambda_{K_b}(x), \tag{10.10}$$

$$\frac{N}{\mu}(\phi, x) = 1 - \phi\Lambda_N(x), \tag{10.11}$$

$$(x = \alpha, \varepsilon, \theta),$$

where μ is rigidity of solid (Pa).

Table 10.1 Mechanical model for solid–liquid aggregates.

Model	Parameters	References
Oblate spheroid	Porosity (ϕ), aspect ratio (α)	Berryman (1980)
Tube	Porosity (ϕ), shape of tube (ε)	Mavko (1980)
Grain shape	Contiguity (φ), porosity (ϕ), wet angle (θ)	Takei (1998)
Crack	Crack density parameter (κ)	O'Connell and Budiansky (1974)

After Takei, Y., 2005. A review of the mechanical properties of solid-liquid composite. Journal of Geography 114 (6), 901–920.

The parameters Λ_{K_b} and Λ_N depend on the shape of pores. If

$$\Lambda_{K_b}(x) = \Lambda_{K_b}(\alpha) \tag{10.12}$$

and

$$\Lambda_N(x) = \Lambda_N(\alpha) \tag{10.13}$$

$$(x = \varepsilon, \theta),$$

then the original model and the oblate spheroid with aspect ratio α give the same K_b and N (Takei, 2002, 2005).

In this condition, we can estimate V_P and V_S using Eqs. (10.1) and (10.2).

10.3 V_P AND V_S INCLUDING LIQUID

V_P/V_S depends on the pore shape (O'Connell and Budiansky, 1974).

If pores are dry, the V_P/V_S decreases with the increase of porosity ϕ. If pores are wet, V_P/V_S increases with the increase of porosity. On the other hand, Watanabe (1993) gives the opposite relation. Takei (2002) explained the opposite conclusions were due to different assumptions on shape of pores.

Takei (2002) gives equations for the case of liquid-filled pores:

$$\frac{\Delta V_P}{V_P^0} = \left[\frac{\dfrac{(\beta-1)\Lambda_{K_b}}{(\beta-1)+\Lambda_{K_b}} + \dfrac{4}{3}\gamma\Lambda_N}{1+\dfrac{4}{3}\gamma} - \left(1 - \frac{\rho_L}{\rho_S}\right) \right] \frac{\phi}{2}, \tag{10.14}$$

$$\frac{\Delta V_S}{V_S^0} = \left[\Lambda_N - \left(1 - \frac{\rho_L}{\rho_S}\right)\right] \frac{\phi}{2}, \tag{10.15}$$

where $\beta = k_s/k_L$ and $\gamma = \mu/k_s$.

β is 5–10, 10–40, and 50–0 for rock and melt, rock and water, and rock and gas, respectively. ρ_L/ρ_S is assumed as 0.92, 0.33 and 0 for melt, water, and gas, respectively.

The ration of V_P decrease and V_S decrease R_{SP} is given by

$$R_{SP} = \frac{\Delta V_S/V_S^0}{\Delta V_P/V_P^0} = \frac{\Lambda_N - \left(1 - \frac{\rho_L}{\rho_S}\right)}{\frac{\frac{(\beta-1)\Lambda_b}{(\beta-1)+\Lambda_b} + \frac{4}{3}\gamma\Lambda_N}{1+\frac{4}{3}\gamma} - \left(1 - \frac{\rho_L}{\rho_S}\right)}. \tag{10.16}$$

According to Takei (2002), V_P/V_S decreases or increases with porosity by $\alpha = 0.1$ or $\alpha = 0.001$, respectively (Fig. 10.1). Thin cracks and dykes given by O'Connell and Budiansky (1974) is the latter case (Takei, 2002).

10.4 EFFECTS OF TEMPERATURE AND PRESSURE

For the enhanced oil recovery (EOR) for heavy oil production, vapor or CO_2 injections have been carried out. By the injection of vapor as in the case of steam-assisted gravity drainage, physical property changes of heavy-oil reservoir have been monitored by V_P and V_S changes. The temperature dependence of heavy oil sands was measured by Nur et al. (1984) and Kato et al. (2008) under confining pressure. Nur et al. (1984) measured the temperature dependence of V_P using Venezuelan oil sand and Kern River sand (Fig. 10.2). Between 15°C and 150°C, the V_P of heavy-oil saturated sand strongly depends on temperature, but V_P of brine or gas-saturated sand does not depend on the temperature. The mixture of oil and brine shows intermediate temperature dependence. Kato et al. (2008) measured the temperature dependence of V_P and V_S of oil shale samples between 10°C and 140°C (Figs. 4.4 and 4.11). They compared the results using Gassmann's equation and found the temperature dependence at temperatures higher than 70°C and 50°C can be explained by Gassmann's equation, but the V_P and V_S showed

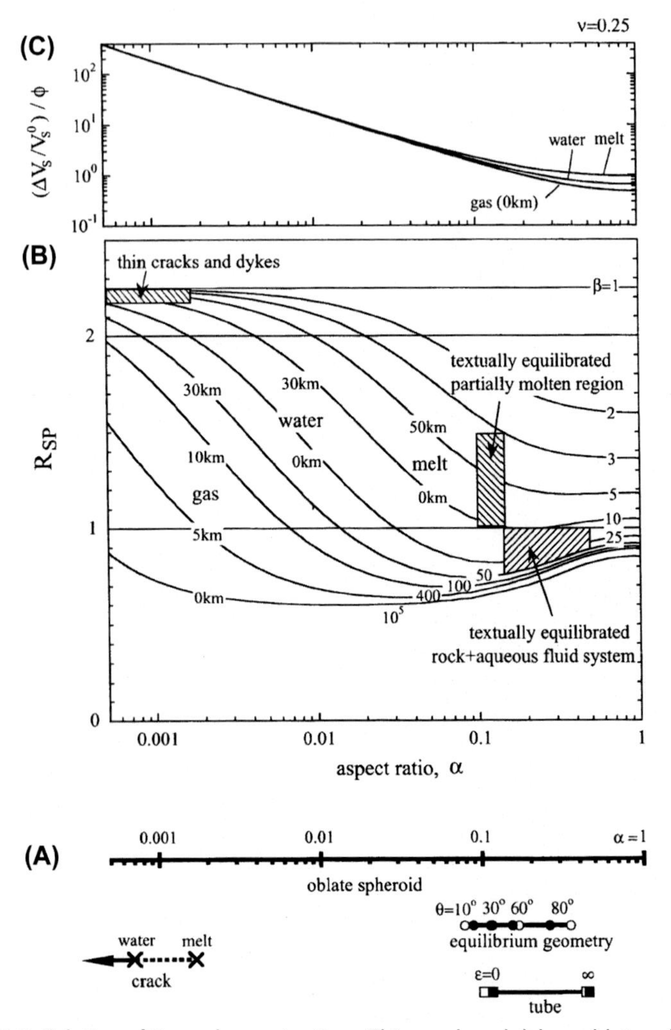

Figure 10.1 Relation of R_{SP} and aspect ratio α. Thin crack and dykes, oblate spheroidal, equilibrium geometry and tube are estimated by aspect ratio α. *(After Takei, Y., 2005. A review of the mechanical properties of solid-liquid composite. Journal of Geography 114 (6), 901–920.)*

higher decrease with temperature than estimation by Gassmann's equation.

The pressure effects are much simpler than temperature dependence. The V_P and V_S increase with pressure increase. There are so many measurements of pressure dependence of V_P and/or V_S

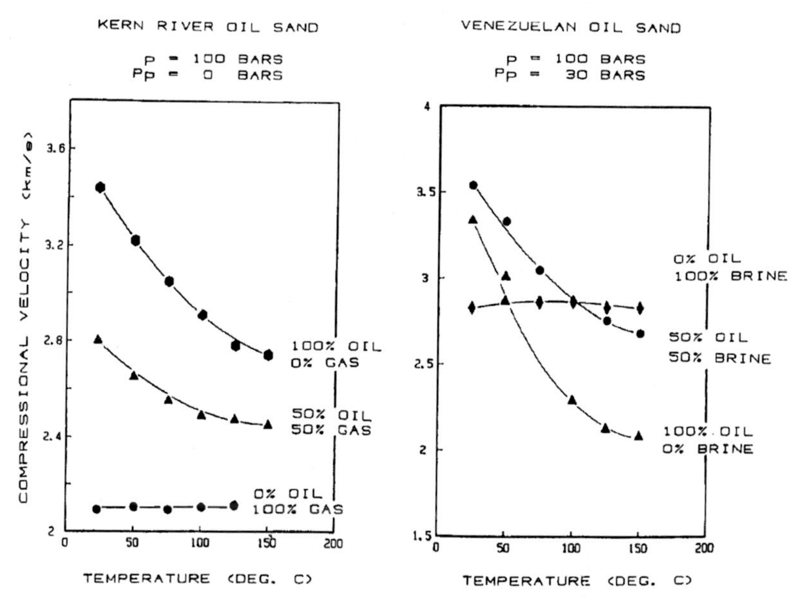

Figure 10.2 Temperature dependence of heavy oil. *(After Nur, A., Tosaya, C., Vo-Thanh, D., 1984. Seismic monitoring of thermal enhanced oil recovery processes: 54th Annual International Meeting, SEG, Expanded Abstracts, 337–340.)*

(Birch, 1960, 1961; Christensen, 1965; Simmons, 1964; Domenico, 1977; Prasad Meissner, 1992; Yin, 1992; Estes et al., 1994; Zimmer, 2003). The process of velocity increase is estimated for the closure of pores by pressure.

10.5 CO_2 INJECTION DURING CARBON CAPTURE AND STORAGE OR CO_2-EOR

The CO_2 injection during carbon capture and storage or CO_2-EOR are discussed and measured in the fields. We evaluated the V_P variation due to the CO_2 saturation using Gassmann's equation (Eqs. (10.6) and (10.8)).

Results are shown in Figs. 10.3 and 10.4 for the gas-phase CO_2 injection and supercritical CO_2 (SCC) injection using the physical properties measured by Hoshino et al. (2001), (Table 10.2). The bulk moduli of water, gaseous CO_2, and SCC used in the calculation were 2.25, 0.006, and 0.046 GPa, respectively (Xue and Ohsumi, 2004). The SCC injection shows much larger velocity change than gas-phase CO_2 injection. In this calculation, the shape of pores was not considered.

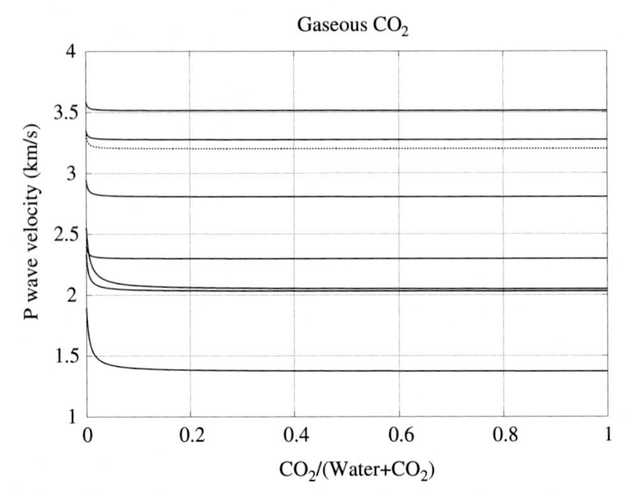

Figure 10.3 The result of Gassmann's estimation of V_P change by gas phase CO_2 injection replacing water phase. Parameters used in this estimation are listed in Table 10.2. Solid lines are mudstone and broken line is volcanic tuff.

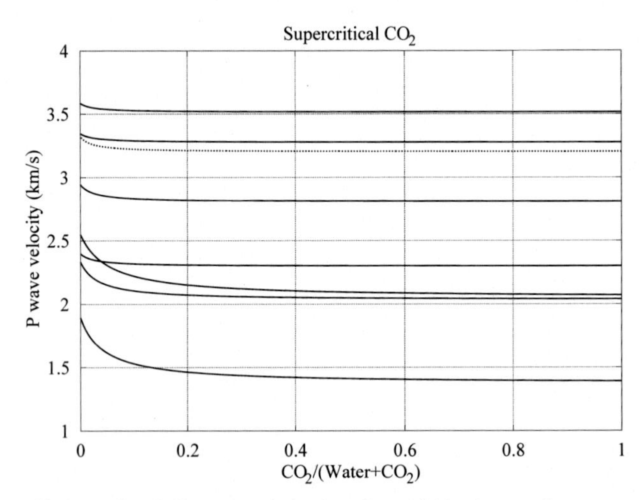

Figure 10.4 The result of Gassmann's estimation of V_P change by supercritical CO_2 injection replacing water phase. Parameters used in this estimation are listed in Table 10.2. Solid lines are mudstone and broken line is volcanic tuff.

Table 10.2 Physical properties used for V_P estimation by CO_2 injection.

No.			K_b (GPa)[a]	k_s (GPa)[b]	Porosity (%)	Density (g/cm^3)	V_S (km/s)
SW0515	Mudstone	Haizume Fm.	1.67	5.5	32.2	1.74	1.54
SW0516	Mudstone	Haizume Fm.	0.824	5.81	28.4	1.91	1.04
SW0517	Tuff	Haizume Fm.	8.22	14.7	28.9	1.77	2.05
SW0518	Mudstone	Haizume Fm.	3.37	5.42	24.3	2.29	1.69
SW0519	Mudstone	Haizume Fm.	0.873	8.82	15.0	2.31	1.69
SW0520	Mudstone	Haizume Fm.	6.87	12.7	20.8	2.33	1.92
SW0521	Mudstone	Haizume Fm.	13.2	18.2	12.6	2.31	2.23
SW0523	Mudstone	Nishiyama Fm.	9.26	12.4	11.05	2.26	2.23

[a]Bulk modulus measured at the confining pressure of 1 kgf/cm^2.
[b]Bulk modulus measured at the confining pressure of 500 kgf/cm^2.
After Hoshino, K., Kato, H., Committee for Compilation of Experimental Data., 2001. Handbook of Mechanical Properties of the Japanese Rocks under High Confining Pressure, Geological Survey of Japan, AIST, 479 p.

Conclusions

This book presents the new Accurately Controlled and Routinely Operated Signals System (ACROSS) methodology in the time-lapse study in oil and gas exploration and production, and carbon capture and storage (CCS).

In the enhanced oil recovery (EOR) technology for the unconventional oil and gas exploration and production, steam or super critical CO_2 is injected to oil and/or gas reservoirs. Injection of steam to tight oil layer could reduce the viscosity of oil to enhance the mobility. Injection of super critical CO_2 could make a miscible state of oil and CO_2, which could also enhance the mobility of oil. The steam-assisted gravity drainage (SAGD) is one of technologies for heavy oil production. The CCS technology aims to reduce the CO_2 emission to the atmosphere by storing CO_2 into subsurface layers.

In shale gas production, injection of fluid to the shale layer could cause the fracking of shale rocks by which natural gas is released from the tight shale. The microearthquake monitoring during fracking is commonly used to monitor the state of fracking shale layers. This is an example of passive seismic time-lapse methods.

In the application of EOR and shale gas fracking technologies during the oil and gas exploration and/or production, understanding the physical state of oil and gas reservoirs and migration of reservoirs is extremely useful and important. If the physical state of oil and gas reservoir is regularly monitored, the efficiency of oil and gas production could be greatly improved. In the CCS technology, the monitoring of injected CO_2 is a mandate to reduce escaping of the injected CO_2 from subsurface to the upper layers and/or atmosphere. The monitoring of physical state and migration of the target reservoirs is called "time-lapse" technology.

Although the time-lapse technology is needed for application of EOR, CCS, and shale gas production by fracking, it seems there are no well-established technologies for the time-lapse method. One of the reasons is that the time lapse is not the extension of ordinary

Time Lapse Approach to Monitoring Oil, Gas, and CO_2 Storage by Seismic Methods
ISBN 978-0-12-803588-7
http://dx.doi.org/10.1016/B978-0-12-803588-7.00011-X
Copyright © 2017
Elsevier Inc.
All rights reserved.

geophysical exploration methods. The 4D seismic method has been widely used for the time lapse because it is the application of the conventional 3D seismic method. However, the costs of the 4D seismic surveys are several times those of 3D seismic surveys. If frequent 4D surveys are needed, the exploration costs are getting high and could be a trade-off with the oil price. In addition, the high repeatability during the 4D seismic exploration is the most difficult factor. The well—well seismic and/or resistivity tomographies are also used for the time lapse, but the region covered by these methods is limited to the area between the wells. The drilling of observation wells also raises the total costs. The vertical seismic profile (VSP) method is less expensive than well—well tomographies but the area of resolved zone is further limited.

There are huge CCS injection regions in the world. Weyburn-Midale in Canada and Sleipner in offshore of Norway are good examples. In these areas the time-lapse methods have been used. Summaries of these CCS time-lapse examples are reviewed in this book. On the other hand, there are some small-scale CCS experimental sites in the world. The Nagaoka CCS test is one of the successful examples. The 10,000 t CO_2 injections were monitored by well—well seismic and resistivity tomographies. The amount of change was approximately 18%. The logging by sonic measurements also showed clear changes due to the CO_2 injection. In Ketzin, Germany, the injection of ~80,000 t CO_2 was observed by 4D seismic method. The simulation done by authors also showed good image of CO_2 injection. However, for the CCS test in the Otway stage-I in Australia, the time-lapse study was not easy.

There are some nonseismic time-lapse methods. Gravity can be used for the time-lapse study. The time lapse using satellite images called Interferometric Synthetic Aperture Radar (InSAR) is widely utilized. The InSAR in In Salah, Algeria, showed clear ground deformation around injection wells.

In this book, the alternative time-lapse technology is presented. The seismic method using ACROSS has several benefits compared to the existing time-lapse methods. A typical seismic source used in ACROSS is a rotational source to generate the centrifugal force by rotation of eccentric mass with an inverter motor well controlled by an accurate time base. One ACROSS seismic source can give 40 tf at 50 Hz. The chirp signal with frequency sweep from a few Hz to 50 Hz during the duration Tw gives a set of line spectra with separation of $1/Tw$. Multiple sweeps are repeated during an hour in a typical ACROSS operation. The seismic waves observed by an array of geophones are treated in frequency domain. The stacking of line spectra corresponding to multiple sweeps in the frequency domain enhances the signal to noise ratio (S/N) by square root of N, where N is number of sweeps. If the one sweep duration is 200 s, 16 sweeps could be stacked during 1 h. One-day data give 19.5 times better S/N of 200 s data. The source signature is well controlled by the feedback system of the inverter motor. The transfer functions between the source and receivers are obtained by division of receiving spectra by source spectra in frequency domain. If needed, the transfer functions can be transformed to time domain by inverse Fourier transformation.

The rotation of source is switched, e.g., every hour. The addition of the transfer functions of clockwise and counterclockwise rotations gives the synthetic response to the vertical vibration if the rotational axis is horizontal. Meanwhile the subtraction of those two transfer functions gives the response to the horizontal vibration. Therefore the transfer functions of vertical and horizontal vibrations are simultaneously obtained by computation. This implies that the seismic data in which P-wave dominates and S-wave dominates can be obtained during, e.g., 2 h.

The repeatability is one of the most important factors in the time-lapse study. There are many factors affecting the repeatability. Uncertainty of source locations leads to the overburden effects. Inaccuracy of a few meters on positioning degrades the normalized root mean square (NRMS) as an index of repeatability. Not only the overburden effects by the inaccuracy of positioning, but the coupling of source with the ground in onshore shooting also causes degradation of NRMS repeatability. The physical property change of near-surface layers is one of the significant factors. The physical properties such as V_P, V_S, attenuation, density, porosity, and anisotropy could be changed by the water contents in the layers and temperature variation. Rainfalls bring large change of water contents in the near-surface layer. The temperature variation at near surface brings frozen of water if temperate goes below the freezing point. These phenomena cause daily and seasonal variations. Even if the borehole geophones are used, the temperature effects and rainfall effects were seen. Precipitation of a few centimeters in a day retains the effects of rainfall for a week. The observed pattern of S-wave arrival time change was well simulated by water flow by precipitation. It is not easy to eliminate the rainfall and temperature effects. The near-surface effect could occur for each of source and receivers. The ACROSS seismic source mounted in the heavy concrete base is used to reduce the source-ground coupling uncertainties.

Through field studies using the ACROSS seismic source in Japan and Saudi Arabia the factors that influence the time lapse are examined. The influence of precipitation and temperature on the repeatability is identified for the geophones placed near surface. By use of buried geophones the repeatability could be greatly improved. In most of previous repeatability tests by many authors it seems difficult to obtain better repeatability than 20% of NRMS. The ACROSS field study in the desert in Saudi Arabia gave very good repeatability such as less than approximately 5% NRMS for two continuous months using an ACROSS seismic source mounted in the concrete block and geophones at the ground surface. These values could include near-surface variation due to the true temporal waveform changes caused by migration of aquifers associated with the water pumping from 64 wells. The temperature change effect could be included in 5%, but it could not be large. If buried geophones at a few tens of meters are used, the NRMS repeatability could be better. In fact, the repeatability of the source itself is estimated as 2% NRMS during 10 days.

The use of a few sources at fixed locations can improve the repeatability in the time-lapse study than those using conventional seismic methods. However, a seismic study using a few sources is quite different from the conventional one. For onshore and offshore surveys 12.5 m or 25 m spacing of receivers and shooting are common. The usefulness of the time lapse using a few sources and an array of geophones is examined by simulations. The back propagation technique is used to get the image of changed zones in the simulations of time lapse. The residual waveforms before and after the temporal change of physical properties in the reservoir area are backpropagated from the receiver positions, then the wave field is expected to focus at the target area.

Using the backpropagation method, several simulations are presented for the heavy oil reservoir cases at 200 m and 2 km depths, the CO_2 storage test field at Ketzin in Germany and the SAGD oil plant in oil-sand in Canada.

For the time-lapse study, hypocenter determination of micro-earthquakes during fracking of oil shale is used. This is one of the passive seismic methods. Compared to this, 4D seismic method, well—well seismic method, VSP method, and ACROSS seismic method are all active seismic methods. The ACROSS methodology can separate the ACROSS signals and background passive signals as natural earthquakes and/or natural or human-made noises. By taking the noise parts and making cross-correlation of noise between two stations we can do seismic interferometry. The results of seismic interferometry using background noise also give the time lapse. If fracking earthquakes in the shale gas production exist during the ACROSS operation, the time lapse by natural earthquakes and active seismic signals can be compared.

The air injection experiment was carried out in Awaji Island, Japan. The 80 tons of air in total was injected to 100 m depth during 5 days. One ACROSS seismic source and 30 geophones were used in the extent of 700 m in EW and 600 m in NS. The residual waveforms showed distinct changes of waveform after the injection. The 12 ms of P-wave arrival time change and the amplitude change were clearly observed after 1 day at 200 m distance. The geophones at 800 m depth with 400 m distance also showed clear waveform changes due to the air injection. By the backpropagation analysis using the residual waveforms the expansion of the influence zone from west to east was clearly imaged. The precipitation effects were identified on the wave amplitudes, but it is not large on travel times.

The clear temporal changes were also identified in the Saudi Arabia test field. Most of large temporal changes could be in surface wave parts and due to the flow of aquifers, while the changes in refracted P-wave arrivals at less than 500 m offset distance are extremely small, say less than 0.2 ms during 2 months for the stations with smallest changes. However some stations show 2 ms changes during 2 months, which probably includes some errors due to the effects of shadow zone by the presence of aquifers below the test field.

Rock physics ruling the temporal change is also discussed.

In conclusion, the time-lapse method is widely used in EOR approach in oil and gas exploration and production, shale gas production, and CCS. As the alternative method of 4D seismic method in the time-lapse approach, the ACROSS methodology with a distinctive seismic source has high potential for the EOR and CCS applications.

APPENDIX A

Fundamentals of Mathematics for ACROSS Processing

A.1 FOURIER TRANSFORM AND DISCRETE FOURIER TRANSFORM

Fourier transform is one of the theoretical backgrounds of ACROSS system. In terms of Fourier series, any time series can be decomposed into periodic oscillation represented by sine and cosine function. The continuous Fourier transform and its inverse transform are defined as

$$X(\omega) = \int_{-\infty}^{\infty} x(t)\exp(-i\omega t)dt, \tag{A.1}$$

$$x(t) = \frac{1}{\pi}\int_{-\infty}^{\infty} X(\omega)\exp(i\omega t)d\omega, \tag{A.2}$$

where t is time, ω is angular frequency, $x(t)$ and $X(\omega)$ are the time series and the frequency series, respectively. The two equations seem quite similar, which means the reciprocity of Fourier transform (see Table A.1).

$$\begin{aligned}
\int_{-\infty}^{\infty} X(t)\exp(-i\omega t)dt &= \int_{-\infty}^{\infty}\left\{\int_{-\infty}^{\infty} x(t')\exp(-it't)dt'\right\}\exp(-i\omega t)dt \\
&= \int_{-\infty}^{\infty} x(t')\left\{\int_{-\infty}^{\infty} \exp(-\,i(\omega + t')t)dt\right\}dt' \\
&= \int_{-\infty}^{\infty} x(t')\delta(\omega + t')dt' \\
&= x(-\omega)
\end{aligned} \tag{A.3}$$

If we assume time series $x(t)$ consists of real numbers, the frequency series $X(\omega)$ generally consists of complex numbers, whose real and imaginary parts represent cosine and sine components,

Table A.1 Main features of continuous and discrete Fourier transform.

	Fourier transform		Discrete Fourier transform	
	Time domain	**Frequency domain**	**Time domain**	**Frequency domain**
Reciprocity	$x(t)$ $X(t)$	$X(\omega)$ $x(-\omega)$	$x(n)$ $X(k)$	$X(k)$ $x(N-n)$
Linearity	$ax(t)+by(t)$	$aX(\omega)+bY(\omega)$	$ax(n)+by(n)$	$aX(k)+bY(k)$
Symmetry	$x(t)\in\mathbb{R}$	$X(-\omega)=\overline{X(\omega)}$	$x(n)\in\mathbb{R}$	$X(N-k)=\overline{X(k)}$
Time shift	$x(t-t_0)$	$X(\omega)\exp(-i\omega t_0)$	$x(n-n_0)$	$X(k)\exp(-2\pi ikn_0/N)$
Frequency shift	$x(t)\exp(i\omega_0 t)$	$X(\omega-\omega_0)$	$x(n)\exp(2\pi ik_0n/N)$	$X(k-k_0)$
Multiplication	$x(t)y(t)$	$X(\omega) * Y(\omega)/\pi$	$x(n)y(n)$	$X(k) * Y(k)$
Convolution	$x(t) * y(t)$	$X(\omega)Y(\omega)$	$x(n) * y(n)$	$NX(k)Y(k)$

respectively. The series of absolute values $\{|X(\omega)|\}$ is called the amplitude spectrum. On the other hand, the series of arguments $\{\mathrm{Arg}\ X(\omega)\}$ is called the phase spectrum.

We usually handle discretely sampled digital time-series data with finite length in actual data processing. Therefore the discrete Fourier transform (DFT) defined as following is used.

$$X(k) = \frac{1}{N}\sum_{n=0}^{N-1} x(n)\exp(-2\pi ikn/N), \quad (k = 0, \cdots, N-1) \tag{A.4}$$

$$x(n) = \sum_{k=0}^{N-1} X(k)\exp(2\pi ikn/N), \quad (n = 0, \cdots, N-1) \tag{A.5}$$

where $X(k)$ is the discrete frequency series that is the Fourier transform of the discrete time series $x(n)$. When the sampling interval and the length of $x(n)$ is Δt and N, respectively, the frequency interval of $X(k)$ is $\Delta f = 1/(N\Delta t)$. $X(0)$ is called DC component with the value of sample mean of the time series data. $X(1)$ is the component of the fundamental frequency $1/(N\Delta t)$ whose period is identical with the whole data length. The following frequency components are considered as the higher harmonics of the fundamental frequency.

If the time series $x(n)$ is real, the frequency components in the second half are reverse-ordered complex conjugate of those in the first half.

$$X(k) = \overline{X(N-k)}, \quad (k = 1, \cdots, N/2 - 1) \tag{A.6}$$

where the overline denotes the complex conjugate. The $N/2$-th frequency with the value of $1/(2\Delta t)$ is called the Nyquist frequency. The signal with the frequency higher than the Nyquist frequency will be aliased, so that antialiasing filters before sampling are used as needed.

Main features of continuous and discrete Fourier transform are summarized in Table A.1.

A.2 WINDOW (BAND LIMIT) EFFECTS

The time window and frequency window are the technique commonly used in digital signal processing. The frequency window has similar meanings with the frequency filter. To extract some range of time or frequency, "window function" is applied.

The simplest window function is a boxcar window, or rectangular window, which is expressed as

$$W_f(\omega) = \begin{cases} 1 & (\omega_1 \le \omega \le \omega_2) \\ 0 & (\omega < \omega_1, \omega_2 < \omega) \end{cases}. \tag{A.7}$$

Considering that the frequency window $W_f(\omega)$ is applied to the frequency series $X(\omega)$, resultant frequency series $Y(\omega)$ is given by multiplication of the two,

$$Y(\omega) = W_f(\omega)X(\omega). \tag{A.8}$$

The inverse Fourier transform of $Y(\omega)$ is

$$y(t) = w_f(t) * x(t), \tag{A.9}$$

where $w_f(t)$ is the Fourier transform of the window function $W_f(\omega)$ and the asterisk denotes the convolution operator defined as

$$W_t(\omega) * X(\omega) = \int_{-\infty}^{\infty} W_t(\omega')X(\omega - \omega')d\omega'. \tag{A.10}$$

The inverse Fourier transform of the boxcar frequency window is derived as

$$\begin{aligned} w_f(t) &= \frac{1}{\pi}\int_{-\infty}^{\infty} W_f(\omega)\exp(i\omega t)d\omega \\ &= \frac{1}{\pi}\int_{\omega_1}^{\omega_2} \exp(i\omega t)d\omega \\ &= \frac{1}{\pi i t}\{\exp(i\omega_2 t) - \exp(i\omega_1 t)\} \\ &= \frac{2}{\pi t}\sin\left(\frac{\omega_2 - \omega_1}{2}t\right)\exp\left(i\frac{\omega_1 + \omega_2}{2}t\right) \\ &= \frac{1}{(\omega_2 - \omega_1)}\text{sinc}\left(\frac{\omega_2 - \omega_1}{2\pi}t\right)\exp\left(i\frac{\omega_1 + \omega_2}{2}t\right). \end{aligned}, \tag{A.11}$$

where sinc(x) is a sinc function or a sampling function defined as

$$\text{sinc}(x) = \frac{\sin \pi x}{\pi x}. \tag{A.12}$$

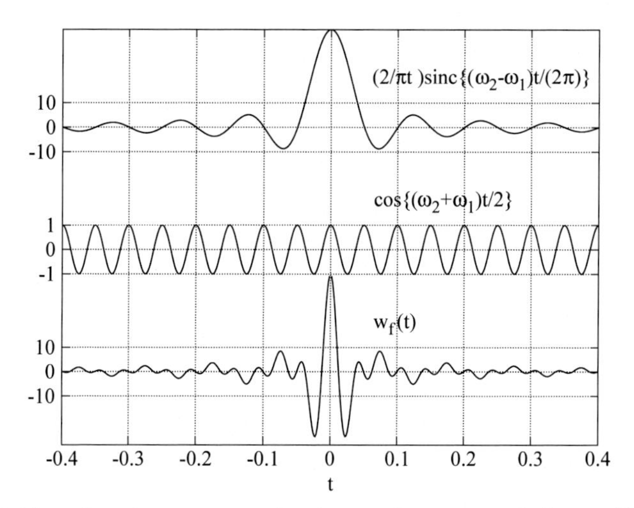

Figure A.1 Time-domain response of a boxcar frequency window $w_f(t)$ (bottom), which is multiplication of sinc function (top) and cosine function (middle).

The window response in Eq. (A.11) is multiplication of sinc function and complex exponential function shown in Fig. A.1. The half width of main lobe of the sinc function is $2\pi/(\omega_2 - \omega_1)$ which means the time resolution is controlled by band width $\omega_2 - \omega_1$ of the frequency window. The exponential part is the frequency shift operator for the center frequency $(\omega_1 + \omega_2)/2$.

In a similar way, a boxcar time window is expressed as

$$w_t(t) = \begin{cases} 1 & (t_1 \leq t \leq t_2) \\ 0 & (t < t_1, t_2 < t) \end{cases}. \tag{A.13}$$

and its Fourier transform is represented by using sinc function, which can be understood by the reciprocity of Fourier transform.

Let us consider another commonly used window function, hanning window,

$$W_f(\omega) = \begin{cases} 0.5 - 0.5\cos 2\pi\dfrac{\omega - \omega_1}{\omega_2 - \omega_1} & (\omega_1 \leq \omega \leq \omega_2) \\ 0 & (\omega < \omega_1, \omega_2 < \omega) \end{cases}. \tag{A.14}$$

Fig. 3.10 in Chapter 3 shows the examples of hanning window and the corresponding time-domain responses. Hanning window has quite small side lobes in comparison to the boxcar window (see Chapter 3).

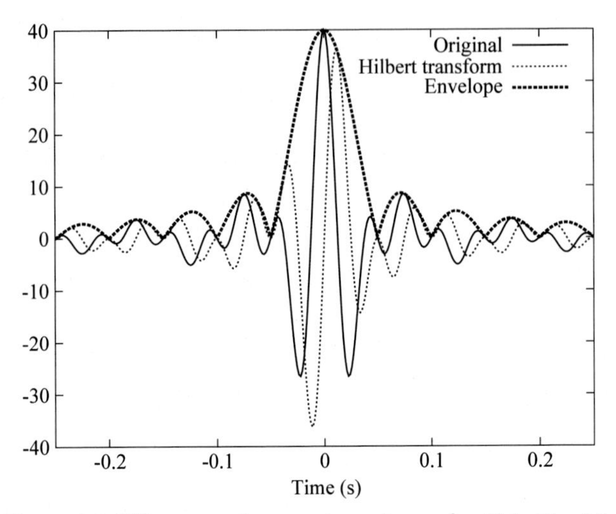

Figure A.2 Hilbert transform and envelope of $w_f(t)$ in Fig. A.1.

A.3 HILBERT TRANSFORM

The Hilbert transform is often used to obtain an envelope of seismograms. The Hilbert transformation of a real time series is equivalent to the filtering in frequency domain

$$H[x(t)] = \mathrm{IFT}[W_H(\omega)X(\omega)], \tag{A.15}$$

where $X(\omega)$ is the Fourier transform of $x(t)$ and the window function $W_H(\omega)$ is

$$W_H(\omega) = \begin{cases} i & (\omega < 0) \\ 0 & (\omega = 0) \\ -i & (\omega > 0) \end{cases}. \tag{A.16}$$

Multiplying $\underline{i}$ means the phase shift of $\pi/2$ so that the Hilbert transform of a cosine function is a sine function. The envelope of the time series $x(t)$ is obtained as

$$\mathrm{Env}[x(t)] = \sqrt{x^2 + H[x(t)]^2}. \tag{A.17}$$

The Hilbert transform and the envelope of $w_f(t)$ in Fig. A.1 are shown in Fig. A.2. The envelope of time-domain response of the boxcar frequency window does not depend on the center frequency but only on the window width.

A.4 FOURIER TRANSFORM OF RANDOM SIGNAL

The concept of random signal is essential for the analysis of natural phenomena. The simplest model of random signal might be a white noise, whose definition is that the average is zero and the values at different times are independent and uncorrelated to each other.

$$E[x(n)] = 0, \quad E[x(n)x(m)] = \sigma^2 \delta_{nm}, \tag{A.18}$$

where $\{x(n)\}$ ($n = 0, \ldots, N$) is a time series of white noise, E is the expectation, σ^2 is the variance, and δ is Kronecker's delta. Due to this definition the autocorrelation of a white noise is the delta function, which is nonzero only at zero lag,

$$p(m) = \sigma^2 \delta_{0m}. \tag{A.19}$$

Therefore the power spectrum $P(k)$, which is the DFT of the autocorrelation, is constant value to σ^2.

$$P(k) = \sigma^2/N, \tag{A.20}$$

where N is the number of points in $\{x(n)\}$. Using DFT defined in (A.4), the power spectrum of a white noise is in inverse proportion to the data length N.

In most cases random noises are assumed to be Gaussian white noise, whose elements independently conform to a normal distribution. Assuming a time series $\{x(n)\}$ of Gaussian white noise conforming to the normal distribution $N(0, \sigma^2)$, the real and imaginary parts of DFT $\{Re[X(k)]\}$, $\{Im[X(k)]\}$ conform to the normal distribution $N(0, \sigma^2/2N)$, where N is the data length. The assumption of Gaussian white noise is the base of noise reduction by stacking.

APPENDIX B

ACROSS Source Details

B.1 PRINCIPLE OF ACROSS

The ACROSS seismic source shown in Fig. 3.4 is a seismic source using the rotation of weight mass with a servomotor (Fig. 3.3). Fig. 3.5 depicts the system of the ACROSS seismic source. The main part of the system consists of a motor with vector inverter made by Mitsubishi Electric Corporation and a signal generator.

The GPS time equipment supplies the accurate time base to the signal generator. The signal generator supplies the necessary pulse sequences to the vector inverter, which controls the position of the weight mass. The pulse sequence is determined by the instantaneous frequency as a function of time within a sweep duration T_m, timing of clockwise and counterclockwise switching, and length of prelude run T_{pr} for the stationary source movement.

The visual interface to the software written by T. Kunitomo is shown in Fig. 3.7. In the case of linear sawtooth frequency modulation, you need to enter the f_{max} and f_{min}, and up:down ratio for each sweep, the period of each sweep, start time, and rotational direction (normal or reverse). The panel seen in Fig. 3.7 shows the frequency and torque histories, current temperatures of lubricant oil and coolant water. In Kunitomo's control program the terminology of frequency modulation carrier frequency f_{ca}, which is the average frequency within a sweep, is used. If the sweep is from 10 to 40 Hz, the carrier frequency is 25 Hz and swing width of frequency width is $\pm$ 15 Hz. This corresponds to $25 \pm 15 = 40$ Hz for f_{max} and $25 - 15 = 10$ Hz for f_{min}.

B.2 SOURCE CONTROL SYSTEM

One rotation of weight mass by the vector invertor motor corresponds to 1 Hz. If we want to generate 50 Hz, the motor rotates

50 times/seconds. The maximum rotation rate of the Mitsubishi inverter motor is 3000 rpm corresponding to 50 rps (Hz). The force generated by rotation of weight mass is given by Eq. (3.9). By increasing the weight of the eccentric mass to get larger centrifugal force, the bearings to support the centrifugal force should be strong enough.

The instantaneous position of the weight mass is controlled by the pulse generated by the signal generator (Figs. 3.5 and 3.6). The rotary encoder used in this vector inverter motor generates 8192 pulses per rotation. In the case of 50 Hz, $50 \times 8192 = 409{,}600$ pulses per second are required. The signal generator gives necessary pulses to control the mass position to designed function of time.

B.3 TIME STANDARD

The time standard is given by the GPS receiver and clock. The accuracy of GPS time is approximately 10 μs. This gives the accuracy of pulses for the position control of the vector inverter motor. Because the position of the weigh mass is controlled by the encoder of motor and the vector inverter, the time accuracy of ACROSS seismic source is less than 10 μs. The time resolution of 0.1 ms can be obtained (see the repeatability section in Chapter 9).

B.4 SWEEP FUNCTION: LINEAR AND NONLINEAR

Frequency sweep is used in the ACROSS seismic source. Three typical sweep patterns of nonlinear, sawtooth, and triangular shape and corresponding spectra are shown in Fig. 3.9. The sweep of triangular shape (upsweep: downsweep = 1:1, Fig. 3.9 right) has strong oscillation of amplitudes in frequency spectra, in comparison to the sawtooth sweep (upsweep: downsweep = 15:1, Fig. 3.9 center). The ratio of upsweep to downsweep controls pattern of spectral amplitude spectra. The larger up:down ratio gives the smoother amplitude spectrum. For the case of the linear sweeps (Fig. 3.9 center and right), the envelope of the amplitude spectra are quadratic functions of frequency, in which amplitudes of low-frequency components are small.

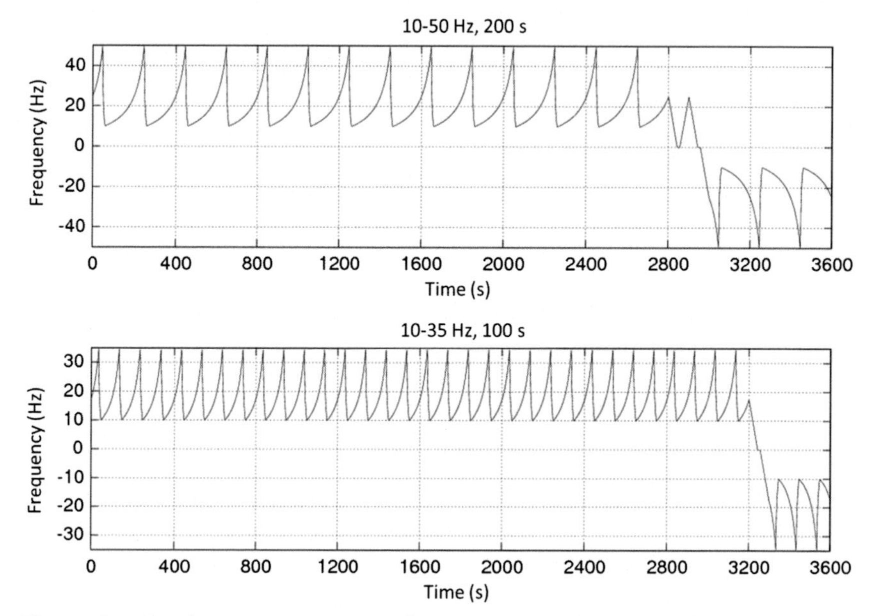

Figure B.1 One-hour sweep patterns for 10—50 Hz with 200 s and 10—35 Hz with 100 s of sweep durations. The rotation changed from normal to reverse at the end of 1 h.

If we need larger force for the low-frequency side, nonlinear sweep as longer sweep-time duration for the low- frequency side and short duration for the high-frequency side is better than the sawtooth or triangular sweep. Although the division of received signals by the source signature in frequency domain can make flat spectra in force, it might lose S/N in low-frequency side. In one case of our ACROSS operations, we used second-order nonlinear sweep from 10 to 50 Hz with 200 s sweep time. Fig. 3.8 shows an example of nonlinear sweep. Small portions of sweeps for 0—1 s and 100—101 s are shown in the middle of Fig. 3.8. Assuming a rotational source as ACROSS, the force and displacement spectra are shown in the bottom of Fig. 3.8.

Two examples of nonlinear sweep pattern during 1 h are shown in Fig. B.1.

B.5 PHASE CONTROL

For the ACROSS seismic source with horizontal rotational axis, the starting position of the weight mass should be controlled to keep at the bottom as 180° from the top. During the prelude time of each switching interval, the mass position is adjusted to the bottom by supplying necessary pulses for position control to the vector inverter. The fluctuation of weigh mass position is ca. a few degrees at 50 Hz. For the low-frequency side the mass position is very accurate.

Using the phase log in the personal computer, we can check the fluctuation of ACROSS sweep control.

B.6 ROTATIONAL DIRECTION

We switched the rotation of ACROSS from clockwise to counterclockwise and vice versa in every hour. However, shorter switching interval might lose efficiency of useable data because each time frame includes the switching time (slowdown–stop–acceleration) and the prelude time for the stationary source movement. You can use longer hours for each rotation direction if necessary.

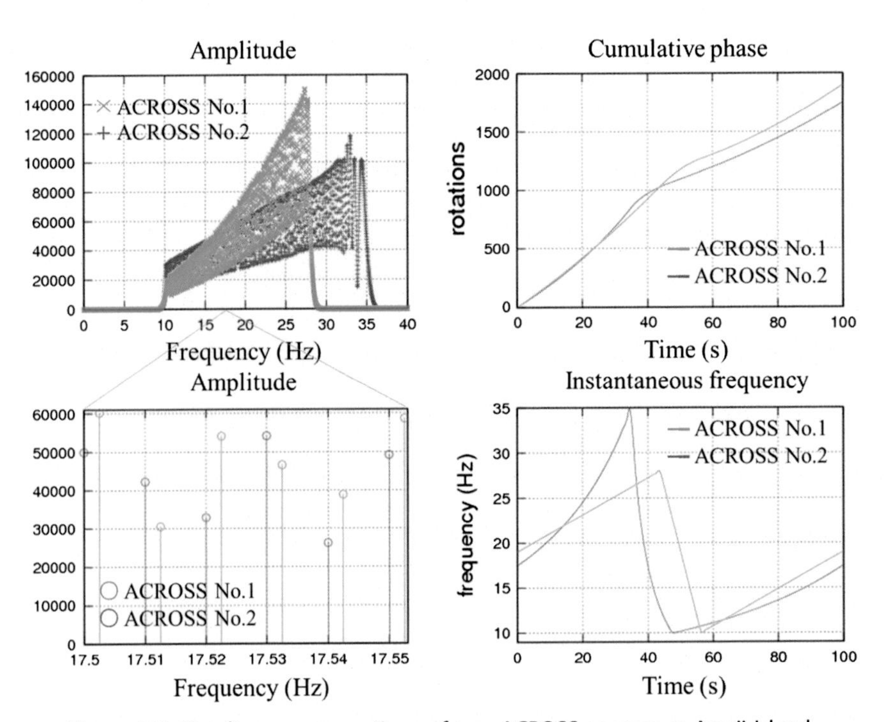

Figure B.2 Simultaneous operations of two ACROSS sources at Awaji Island.

B.7 SIMULTANEOUS OPERATION OF PLURAL ACROSS SOURCES

Without using separate frequency bands, we can operate plural ACROSS seismic sources simultaneously. Fig. B.2 is an example of two sources' operations at the same time. The line spectra of the signals from two source units are completely distinct and independent shaping interdigitating comb teeth.

APPENDIX C

Processing of Acquired Data

C.1 DISCRETE FOURIER TRANSFORM

The first step of processing ACROSS data is discrete Fourier transform (DFT). Details about DFT and Fourier transform are described in Appendix A.1. DFT converts finite discrete time series data into finite discrete frequency spectra. The main parts of overall processing are done in frequency domain.

Original data acquired by ACROSS system are time series of the ground motion observed by seismometers. Sampling clock of the recorder should be synchronized to GPS time base. Before applying DFT, the observed time series must be segmented in accordance with the source operation. Key parameters of the source operation are the repetition period T, the carrier frequency f_c, the frequency range $[f_1, f_2]$ and the period of switching rotation direction T_s. The repetition period T must be an integer multiple of the sampling rate Δt of the recorders. Observed time series data are segmented by the length of mT, an integer multiple of T. While the repetition period T is the recurrence time of the frequency modulation, the segment length mT must be the integer multiple of the recurrence period in the strict meaning including the phase or the position of eccentric mass.

Assuming a linear saw tooth modulation with $f_1 = 10.05$ Hz, $f_2 = 20.05$ Hz, and sweep period of $T = 10$ s, the carrier frequency or the average frequency f_c is 15.05 Hz. The rotation number in 10 s is 150.5, which means the mass position after 10 s is 180° from the original position. In this case, the strict recurrence time is 20 s. It is important that $f_c mT$, the total rotation number in mT, must be an integer.

Fig. C.1 shows an example of frequency transition of rotary ACROSS source.

DFT is applied to each segment just as the length mT without any tapers or zero paddings. Here we use the definition of DFT,

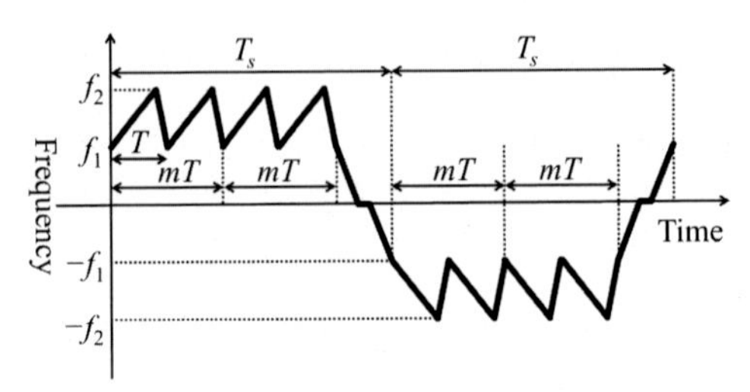

Figure C.1 A schematic of frequency transition of rotational ACROSS source. Negative frequency means rotation in the opposite direction. In this example, rotation direction is switched every period T_s; within it the unit sweep T is repeated four times. The segment length for DFT is $mT = 2T$.

$$X(k) = \frac{1}{N}\sum_{n=0}^{N-1} x(n)\exp(-2\pi ikn/N), \quad (k = 0, \cdots, N-1) \qquad \text{(C.1)}$$

The resultant $X(k)$ is a frequency series with the frequency spacing of $1/mT$, the corresponding frequency $f(k)$ is $k/(mT)$.

C.2 SEPARATION OF NOISE SPECTRA FROM SIGNAL SPECTRA

The result of DFT (Eq. C.1) is a set of discrete complex series $\{X(k)\}$ of the spectrum, whose frequency spacing is $1/(mT)$. The signal with the frequency modulation period T and the carrier frequency f_c is composed of discrete line spectra around f_c with the spacing of $1/T$. If the ACROSS source operation is precisely cyclic and synchronized to the sampling timing, the source signals exist at discrete frequencies that we exactly know. Details about calculation of the source spectrum will be mentioned in the later section.

Thus the ACROSS-signal components can be extracted by picking data of the frequencies that source signal exists. The other frequency components do not contain the ACROSS signal but consist of noises if the instantaneous frequency of ACROSS signal precisely repeats in the cycle T.

The discrete feature of the ACROSS signal in frequency domain enables us to operate plural sources with the same frequency range simultaneously and observe the response for each source and also background noise independently.

The frequency components excluding the ACROSS signal consist of background noises if the ACROSS source is completely controlled and the temporal change of surroundings is negligible. Assuming that the amplitude spectrum of the background noise is smooth, each spectral amplitude of noise at the signal frequency is interpolated as the root mean square of amplitude spectra at neighboring noise frequencies.

C.3 ENHANCEMENT OF S/N RATIO BY STACKING

Stacking or averaging is generally used to enhance signal-to-noise ratio (S/N), which is based on randomness of the background noise. Nagao et al. (2010) has developed the statistically optimum stacking method for ACROSS system. Through the process described in the previous section, the signal spectra and interpolated noise amplitude spectra are obtained for every signal frequency. The stacking method uses the noise amplitude for calculating the weight. The weight for the frequency f of the k-th time segment is

$$w(f,k) = \frac{1}{\varepsilon(f,k)^2} \Big/ \sum_{k'=1}^{K} \varepsilon(f,k')^2, \tag{C.2}$$

where $\varepsilon(f, k)$ is estimated noise level. The stacked spectrum and its error are

$$\begin{aligned} X_S(f) &= \sum_{k=1}^{K} w(f,k)X(f,k), \\ \varepsilon_S(f) &= \left\{ \sum_{k=1}^{K} \varepsilon(f,k)^2 \right\}^{-1/2}. \end{aligned} \tag{C.3}$$

C.4 CALCULATION OF SOURCE SPECTRUM

One of the distinctive features of ACROSS system is that the source signal is known. It is ideal to know the actual time function of the force excited by the source as precise as possible. In the case of rotary-type source, centrifugal force excited by the eccentric mass is considered. The centrifugal force vector $\mathbf{f}(t)$ is given by

$$\mathbf{f}(t) = -M\frac{d^2\mathbf{r}(t)}{dt^2}, \tag{C.4}$$

where $\mathbf{r}(t)$ is the position vector of the center of gravity of the mass M.

There are roughly two ways to obtain the mass movement. One is to measure the mass position using rotary encoder and the other is to use the design time function. The count of pulses from rotary encoder gives the cumulative rotation angle $\theta(t)$. The design function of instantaneous frequency of frequency modulation (FM) signal is integrated to get the cumulative rotation angle.

Now we set the coordinate (X, Y, Z) for rotary-type source as shown in Fig. C.2. X is the direction that the mass locates at the reference time $t=0$, Y is the direction rotated 90° from X, and Z is the direction of rotation axis. The mass position vector is therefore represented using the cumulative rotation angle as

$$\mathbf{r}(t) = [R\cos\theta(t), R\sin\theta(t), 0]^{\mathrm{T}}, \tag{C.5}$$

where R is the rotation radius of the mass and the superscript T denotes the transpose.

The Fourier transform of $\mathbf{r}(t)$ is

$$\mathbf{R}(\omega) = R[C(\omega), S(\omega), 0]^{\mathrm{T}}, \tag{C.6}$$

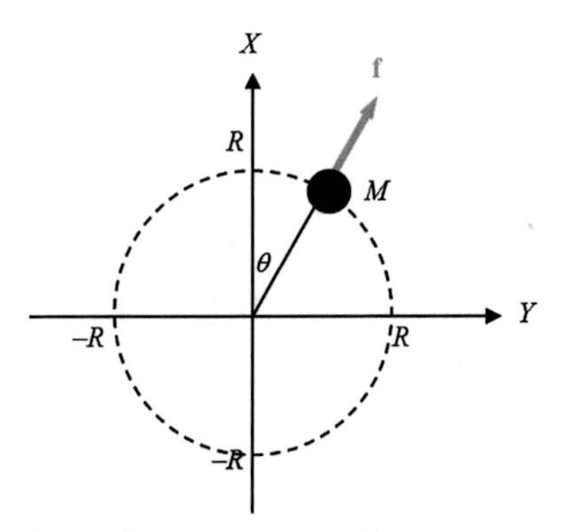

Figure C.2 Schematic of centrifugal force excited by rotating mass and the coordinate system. X is the direction of the original mass position, Y is the direction rotated 90° from X, and Z is rotation axis perpendicular to the paper.

where $C(\omega)$ and $S(\omega)$ are the Fourier transform of $\cos\theta(t)$ and $\sin\theta(t)$, respectively. Then the force spectrum is obtained by

$$\mathbf{F}(\omega) = -M(i\omega)^2\mathbf{R}(\omega) = MR\omega^2[C(\omega), S(\omega), 0]^{\mathrm{T}} \tag{C.7}$$

In actual processing, the DFTs $C(k)$ and $S(k)$ are calculated using the discrete sampled series of cumulative rotation angle $\theta(n)$

$$\begin{aligned} C(k) &= \frac{1}{N}\sum_{n=0}^{N-1}\cos\theta(n)\exp(-2\pi ikn/N), \\ S(k) &= \frac{1}{N}\sum_{n=0}^{N-1}\sin\theta(n)\exp(-2\pi ikn/N), \quad (k = 0, \cdots, N-1). \end{aligned} \tag{C.8}$$

The length N of $\theta(n)$ must be the repetition period of the mass position, not the period of frequency modulation. The force spectra in two directions are calculated by Eq. (C.7).

C.5 SOURCE SIGNATURE DECONVOLUTION: TRANSFER FUNCTION

Assuming a linear system, the spectrum of observed ground motion is multiplication of the source spectrum and the transfer function as mentioned. If both observed and source spectra are scalar, the transfer function can be obtained by simple division:

$$H(\omega) = \frac{U(\omega)}{F(\omega)}, \tag{C.9}$$

where $H(\omega)$, $U(\omega)$, and $F(\omega)$ are complex frequency series of the transfer function, the observed spectrum, and the source spectrum, respectively. Note that zero amplitude $F(\omega)$ makes $H(\omega)$ diverge to infinity. Calculation should be made for the signal frequency of the ACROSS source.

As mentioned in Section C.3, the estimated error of observed spectrum is also obtained. Therefore the estimated error of the transfer function can be calculated by error propagation:

$$E^2(\omega) = \frac{\varepsilon^2(\omega)}{|F(\omega)|^2}, \tag{C.10}$$

where $E(\omega)$ and $\varepsilon(\omega)$ are estimated errors of the transfer function and the observed spectrum, respectively.

If the force at the source is single force in one direction, the calculation becomes similar to the scalar case discussed earlier. However, because the rotary-type ACROSS source excites two-dimensional force as Eq. (C.7), it is required the vector and matrix calculation presented in the following section.

C.6 CONVERSION FROM ROTATION TO SYNTHETIC LINEAR VIBRATION

In general, the transfer function of the three-component vector displacement for the three-component vector force is a second-order tensor with nine components:

$$\mathbf{U}(\omega) = \mathbf{H}(\omega)\mathbf{F}(\omega),$$
$$\begin{bmatrix} U_x(\omega) \\ U_y(\omega) \\ U_z(\omega) \end{bmatrix} = \begin{bmatrix} H_{xX}(\omega) & H_{xY}(\omega) & H_{xZ}(\omega) \\ H_{yX}(\omega) & H_{yY}(\omega) & H_{yZ}(\omega) \\ H_{zX}(\omega) & H_{zY}(\omega) & H_{zZ}(\omega) \end{bmatrix} \begin{bmatrix} F_X(\omega) \\ F_Y(\omega) \\ F_Z(\omega) \end{bmatrix}. \tag{C.11}$$

If we have observations for three linearly independent vector forces, the simultaneous equations for the tensor transfer function can be solved.

$$\mathbf{H}(\omega) = [\mathbf{U}_1(\omega) \quad \mathbf{U}_2(\omega) \quad \mathbf{U}_3(\omega)][\mathbf{F}_1(\omega) \quad \mathbf{F}_2(\omega) \quad \mathbf{F}_3(\omega)]^{-1}. \tag{C.12}$$

Because the force excited by rotational source does not have Z-component as shown in Eq. (C.7), we consider six components of the transfer function:

$$\begin{bmatrix} U_x(\omega) \\ U_y(\omega) \\ U_z(\omega) \end{bmatrix} = \begin{bmatrix} H_{xX}(\omega) & H_{xY}(\omega) \\ H_{yX}(\omega) & H_{yY}(\omega) \\ H_{zX}(\omega) & H_{zY}(\omega) \end{bmatrix} \begin{bmatrix} F_X(\omega) \\ F_Y(\omega) \end{bmatrix}. \tag{C.13}$$

In this case, the observations for two linearly independent vector forces provide solution of the six-component transfer function.

$$\begin{bmatrix} H_{xX}(\omega) & H_{xY}(\omega) \\ H_{yX}(\omega) & H_{yY}(\omega) \\ H_{zX}(\omega) & H_{zY}(\omega) \end{bmatrix} = \begin{bmatrix} U_{1x}(\omega) & U_{2x}(\omega) \\ U_{1y}(\omega) & U_{2y}(\omega) \\ U_{1z}(\omega) & U_{2z}(\omega) \end{bmatrix} \begin{bmatrix} F_{1X}(\omega) & F_{2X}(\omega) \\ F_{1Y}(\omega) & F_{2Y}(\omega) \end{bmatrix}^{-1}. \tag{C.14}$$

Substituting Eqs. (C.7)−(C.14),

$$\begin{bmatrix} H_{xX}(\omega) & H_{xY}(\omega) \\ H_{yX}(\omega) & H_{yY}(\omega) \\ H_{zX}(\omega) & H_{zY}(\omega) \end{bmatrix} = [\mathbf{U}_{+}(\omega) \quad \mathbf{U}_{-}(\omega)] \frac{1}{MR\omega^2} \begin{bmatrix} C(\omega) & C(\omega) \\ S(\omega) & -S\omega) \end{bmatrix}^{-1}, \tag{C.15}$$

where $\mathbf{U}_{+}(\omega)$ and $\mathbf{U}_{-}(\omega)$ are the observed displacement for clockwise and counterclockwise rotations, respectively. The force vector for counterclockwise rotation is expressed by the reverse sign in Y-component. This equation can be simplified as

$$\begin{bmatrix} H_{xX}(\omega) & H_{xY}(\omega) \\ H_{yX}(\omega) & H_{yY}(\omega) \\ H_{zX}(\omega) & H_{zY}(\omega) \end{bmatrix} = \left[\frac{\mathbf{U}_{+}(\omega) + \mathbf{U}_{-}(\omega)}{2MR\omega^2 C(\omega)} \quad \frac{\mathbf{U}_{+}(\omega) - \mathbf{U}_{-}(\omega)}{2MR\omega^2 S(\omega)}\right], \tag{C.16}$$

where the transfer function components for the X and Y forces are obtained by addition and subtraction of the displacements by clockwise and counterclockwise rotation.

C.7 TIME-DOMAIN WAVEFORM

The fundamental observable of ACROSS system is the transfer function in frequency domain, which can be obtained through the above-mentioned process. However, the time-domain waveform is often required for further analysis and interpretation. The time-domain waveform of a transfer function, which is referred to as the impulse response function, is given by the inverse Fourier transformation.

In the case of ACROSS measurement, the transfer functions are only obtained for discrete frequencies in limited range. So we can only obtain the band-limited version of the impulse response function,

$$h(t) = \sum_{j=1}^{N} H(f_j)\exp(2\pi i f_j t), \tag{C.17}$$

where $\{H(f_j)\}$ is the transfer function for discrete frequencies $f_1, \ldots, f_N$.

For further analysis and interpretation, one should be aware of the effect of frequency window, especially time resolution and side lobes.

REFERENCES

Akaike, H., 1969a. Fitting autoregressive models for prediction. Annals of the Institute of Statistical Mathematics 21, 243–247.

Akaike, H., 1969b. Power spectral estimation through autoregressive model fitting. Annals of the Institute of Statistical Mathematics 21, 407–419.

Arts, R., Eiken, O., Chadwick, R.A., Zweigel, P., van der Meer, L., Zinszner, B., 2004. Monitoring of CO_2 injected at Sleipner using time-lapse seismic data. Energy 29 (9–10), 1383–1392.

Arts, R., Chadwick, R.A., Eiken, O., Thibeau, S., Nooner, S., 2008. Ten years experience of monitoring CO_2 injetion in the Utsira Sand at Sleipner, offshore Norway. First Break 26 (1), 65–72.

Barkved, O., 2011. Valhall PRM-Technical highlights from strength to strength. In: Extended Abstract of EAGE Workshop on PRM (Permanent Reservoir Monitoring) Using Seismic Data, pp. 20–23 (Tronheim, Norway).

Berryman, J.G., 1980. Long-wavelength propagation in composite elastic media 1. Elliptical inclusions. The Journal of the Acoustical Society of America 68, 1820–1831.

Biot, M.A., 1956a. Theory of propagation of elastic waves in a fluid-saturated porous solid, 1, low-frequecy range. The Journal of the Acoustical Society of America 28, 168–178.

Biot, M.A., 1956b. Theory of propagation of elastic waves in a fluid-saturated porous solid, 2, higher-frequency range. The Journal of the Acoustical Society of America 28, 179–191.

Birch, F., 1960. The velocity of compressional waves in rocks to 10 kilobars, 1. Journal of Geophysical Research 65, 1083–1102.

Birch, F., 1961. The velocity of compressional waves in rocks to 10 kilobars, 2. Journal of Geophysical Research 66, 2199–2224.

Bishop, J.E., Davis, T.L., 2014. Multi-component seismic monitoring of CO_2 injection at Delhi Field, Louisiana. First Break 32 (5), 43–48.

Burg, J.P., 1967. Maximum Entropy Spectral Analysis: 37th Annual International Meeting. SEG.

Burg, J.P., 1968. A New Analysis Technique for Time Series Data, Paper Presented at Advance Study Institute on Signal Processing. NATO, Enscheda, Netherlands.

Calvert, R., 2005. Insight and Method for 4D Reservoir Monitoring and Characterization. SEG.

Cantillo, J., 2011. A Quantitative Discussion on Time-lapse Repeatability and Metrics: 81st Annual International Meeting. SEG, pp. 4160–4164. Expanded Abstracts. http://dx.doi.org/10.1190/1.3628075.

Chadwick, R.A., Noy, D., Arts, R., Eiken, O., 2009. Latest time-lapse seismic data from Sleipner yield new insight CO_2 plume development. In: 9th International Conference on Greenhouse Gas Control Technology, Washington D.C. USA, 17–20, November 2008. ISSN: 18766102. Elsevier (Pub).

Chadwick, A., Williams, G., Delepine, N., Clochard, V., Labat, K., Sturton, S., Buddensiek, M.L., Dillen, M., Nickel, M., Lima, A.L., Arts, R., Neele, F., Rossi, G., 2010. Monitoring Quantitative analysis of time-lapse monitoring data at the Sleipner CO_2 storage operation. The Leading Edge 29, 170–177.

Chen, H.F., Li, X.Y., Qian, Z.P., Zhao, G.L., 2013. Robust adaptive polarization analysis method for eliminating ground roll in 3C land seismics. Applied Geophysics 10 (3), 295–304.

Christensen, N., 1965. Compressional wave velocities in metamorphic rocks at pressures to 10 kilobars. Journal of Geophysical Research 70, 6147–6164.

Claerbout, J.F., 1968. Synthesis of a layered medium from its acoustic transmission response. Geophysics 33, 264–269.

Claerbout, J.F., 1992. Earth Soundings Analysis: Processing versus Inversion. Blackwell Scientific Publication.

Clochard, V., Delepine, N., Labat, K., Ricate, P., 2010. CO_2 plume imaging using 3D pre-stack stratigraphic inversion: a case study on the Sleipner field. First Break 28 (1), 91–96.

Davis, T.L., Terrell, M.J., Benson, R.D., Cardona, R., Kendall, R., Winarsky, R., 2003. Multi-component seismic characterization and monitoring CO_2 flood at Weyburn field, Saskatchewan. The Leading Edge 22 (7), 696–697.

Davis, T., 2015. Time-lapse multi-component seismic monitoring, Delhi field, Louisiana. First Break 33 (2), 65–70.

Domenico, S.N., 1977. Elastic properties of unconsolidated porous sand reservoirs. Geophysics 42, 1339–1368.

De Jongh, S., 2011. Valhall-imapact of frequent 4D seismic on field delivery. In: Extended Abstract of EAGE Workshop on PRM (Permanent Reservoir Monitoring) Using Seismic Data, pp. 154–157 (Tronheim, Norway).

Eiken, O., Haugen, G.U., Schonewille, M., Duijndam, A., 2003. A proven method for acquiring highly repeatable towed streamer seismic data. Geophysics 68, 1303–1309.

Estes, C., Mavko, A., Yin, G., Cadoret, H., 1994. Measurements of velocity, porosity, and permeability on unconsolidated granular materials: Stanford Rock Physics and Borehole Geophysical project. Annual Reports 55. G1–G1-G1909.

Folstad, P.G., 2011. Ekofisk-Justification for permanent monitoring system. In: Extended Abstract of EAGE Workshop on PRM (Permanent Reservoir Monitoring) Using Seismic Data, pp. 142–145 (Trondheim, Norway).

Furre, A.-K., Eiken, O., 2014. Dual sensor streamer technology used in Sleipner CO2 injection monitoring. Geophysical Prospecting 62, 1075–1088.

Gassmann, F., 1961. Über die elastizität poröser medien: Vierteljahrsschrift der Naturforschenden Gesellschaft in Zurich, vol. 96, pp. 1–23.

Ghaderi, A., Landrø, L., 2009. Estimation of thickness and velocity changes of injected carbon dioxide layers from prestack time-lapse seismic data. Geophysics 74 (2), O17–O28.

Hill, R., 1952. The elastic behavior of crystalline aggregates. Proceedings of the Physical Society of London Section A 65, 349.

Hoshino, K., Kato, H., Committee for Compilation of Experimental Data, 2001. Handbook of Mechanical Properties of the Japanese Rocks under High Confining Pressure, Geological Survey of Japan. AIST, 479 p.

Ivanova, A., Kashubin, A., Juhojuntti, N., Kummerow, J., Henninges, J., Juhlin, C., Lüth, S., Ivandic, M., 2012. Monitoring and volumetric estimation of injected CO_2 using 4D seismic, petrophysical data, core measurements and well logging: a case study at Ketzin, Germany. Geophysical Prospecting 60, 957–973.

Ivandic, M., Yang, C., Lüth, S., Cosma, C., Juhlin, C., 2012. Time-lapse analysis of sparse 3D data from the CO_2 storage pilot site at Ketzin, Germany. Journal of Applied Geophysics 84, 14–28.

Jervis, M., Bakulin, A., Bursstad, R., Berron, C., Fogues, E., 2012. In: Suitability of Vibrator for the Time-lapse Monitoring in the Middle East, SEG Las Vegs, 2012, Annual Meeting. http://dx.doi.org/10.1190/segam2012-0948.1.

Johnston, D., 2013. Practical Applications for Time-lapse Seismic Data, 2013 Distinguished Instructor Short Course, Distinguished Instructor Series, No. 16. SEG, 270 pp.

Juhlin, C., Giese, R., Zinck-Jørgensen, K., Cosma, C., Kazemeini, H., Juhojuntti, N., Lüth, S., Norden, B., Förster, A., 2007. 3D baseline seismics at Ketzin, Germany: the CO2SINK project. Geophysics 72, B121–B132.

Kasahara, J., Tomoda, Y., 1993. An Introduction to Computerized Geoscience. University Tokyo Press, p. 239.

Kasahara, J., Korneev, V., Zhdanov, M. (Eds.), 2010a. Active geophysical monitoring. Handbook of Geophysical Exploration, Seismic Exploration, vol. 40. Elsevier Pub, p. 551.

Kasahara, J., Hasada, Y., Tsuruga, K., 2010b. Seismic imaging of time lapse for CCS and oil and gas reservoirs using ultra-stable seismic source (ACROSS). In: Proceedings of the 123rd SEGJ Conference, pp. 74–77.

Kasahara, J., Hasada, Y., Tsuruga, K., 2011a. Imaging of ultra-long term temporal change of reservoir(s) by accurate seismic sources(s) and multi-receivers. In: Extended Abstract of EAGE Workshop on Permanent Reservoir Monitoring (PRM) Using Seismic Data, pp. 40–44 (Trondheim, Norway).

Kasahara, J., Ito, S., Hasada, Y., Takano, M., Guidi, A., Fujiwara, T., Tsuruga, K., Fujii, N., 2011b. Time lapse experiment using the seismic ACROSS source near the Nojima-fault in Awaji Island- Generation of vertical and horizontal vibration by synthetic method. In: Proceedings of the 125th SEGJ Conference, pp. 151–154.

Kasahara, J., Yamaishi, T., Ito, S., Fujii, N., Nakagawa, I., Hasada, Y., Yamaoka, K., Ikuta, R., Nishigami, K., 2011d. Injected air diffusion and influence of rain fall in the near surface ground near the Nojima Fault in Awaji Island. In: Proceedings of the 125th SEGJ Conference, pp. 163–166.

Kasahara, J., Hasada, Y., Yamaoka, K., Ikuta, R., 2012. Imaging of the fluid injection zone at the Nojima-fault: reanalysis of 2003 Awaji Island water injection experiment. In: Japan Geoscience Union Annual Meeting Abstract.

Kasahara, J., Kato, A., Takanashi, M., Hasada, Y., Lüth, S., Juhlin, C., 2013a. Simulation of time-lapse for the Ketzin) CO_2 storage site assuming a single seismic ACROSS and multi-seismic receivers. In: EAGE Annual Meeting Extended Abstract.

Kasahara, J., Kato, A., Takanashi, M., Hasada, Y., 2013b. Time lapse simulation for SAGD (Steam-Assisted Gravity Drainage) of JACOS (Japan Canada Oil Sands) assuming seismic ACROSS and 2D geophone array. In: Proceedings of the 128th SEGJ Conference, pp. 103–105.

Kasahara, J., Ito, S., Fujiwara, T., Hasada, Y., Tsuruga, K., Ikuta, R., Fujii, N., Yamaoka, K., Ito, K., Nishigami, K., 2013c. Real time imaging of CO_2 storage zone by very accurate stable-long term seismic source. Energy Procedia 37, 4085–4092.

Kasahara, J., Kubota, R., Kanai, Y., Tazawa, O., Fujimoto, O., Nishiyama, E., Kamimura, A., Murase, K., Noguchi, S., Ohmura, T., Hasada, Y., 2014a. A time lapse test of seismic waveform changes during several days at a tuff area in Japan using a seismic vibrator. In: 2nd International KAUST-KACST-JCCP Workshop on Surface and Subsurface 4D Monitoring, King Abdullah University of Science and Technology (KAUST), Saudi Arabia, 4–6 March.

Kasahara, J., Kubota, R., Kanai, Y., Tazawa, O., Fujimoto, O., Nishiyama, E., Kamimura, A., Murase, K., Noguchi, S., Ohmura, T., Hasada, Y., 2014b. Continuous monitoring of seismic waveform changes during several days at the green tuff area in Japan using a seismic vibrator. In: Proceedings of the 8th Asian Rock Mechanics Symposium, PE1-4, pp. 1615–1624 (Sapporo, Japan).

Kasahara, J., Hasada, Y., Al-Damegh, K., Alanezi, G.T., Nishiyama, E., 2014c. Seismic interferometry using the data during ACROSS operation. In: Proceedings of the 131th SEGJ Conference, pp. 147–149.

Kasahara, J., Hasada, Y., Kamimura, A., Fujii, N., Ushiyama, M., 2015a. Source signature changes using conventional seismic source. In: Proceedings of the 132th SEGJ Conference, pp. 203–206.

Kasahara, J., Fujii, N., Hasada, Y., Kamimura, A., Ushiyama, M., 2015b. Seismic time lapse experiments in the old quarry and the relation to the past subsidence area. In: Proceedings of the 132th SEGJ Conference, pp. 207–210.

Kasahara, J., Al Damegh, K., Alanezi, G., AlYousef, K., Almalki, F., Lafouza, O., Hasada, Y., Murase, K., 2015c. Simultaneous time-lapse data acquisition of active and passive seismic sources at Al Wasse water pumping field in Saudi Arabia. Energy Procedia 76, 512–518.

Kasahara, J., Al Damegh, K., Al-Anezi, G., Murase, K., Kamimura, A., Fujimoto, O., Ohnuma, H., Hasada, Y., 2016a. Time-lapse observation in Al Wasse field in Saudi Arabia using ACROSS seismic source and its interpretation. In: Proceedings of the 134th SEGJ Conference, pp. 104–107.

Kasahara, J., Al Damegh, K., Al-Anezi, G., Murase, K., Kamimura, A., Fujimoto, O., Ohnuma, H., Hasada, H., 2016b. ACROSS Refraction study for the interpretation of time-lapse data in Al Wasse field, Saudi Arabia. In: Proceedings of the 134th SEGJ Conference, pp. 100–103.

Kasahara, J., Al Damegh, K., Al-Anezi, G., Hasada, Y., Murase, K., Kamimura, A., Fujimoto, O., Ohnuma, H., 2016c. Repeatability estimation of ACROSS in Al Wasse field, Saudi Arabia. In: Proceedings of the 135th SEGJ Conference.

Kato, A., Onozuka, S., Nakayama, T., 2008. Elastic property changes in a bitumen reservoir during steam injection. The Leading Edge 27 (9), 1124–1131.

Kinkela, J., Pevzner, R., Urosevic, M., 2011. Ground roll repeatability analysis – CO2CRC Otway project case study. In: 73rd EAGE Conference & Exhibition – Workshops, p. I039.

Koottungal, L., April 2, 2012. World EOR survey. Oil & Gas Journal 57–69.

Koottungal, L., April 9, 2014. 2014 Worldwide EOR survey. Oil & Gas Journal 1–15.

Kragh, E., Chirstie, P., July 2002. Seismic repeatability, normalized rms and predictability. The Leading Edge 21 (7), 640–647.

Kristiansen, P., Christie, P., Bouska, J., O'Donovan, A., Westwater, P., 2000. Foinaven 4D: Processing and Analysis of Two Designer 4Ds. SEG 2000 Expanded Abstracts.

Kubota, R., Kanai, Y., Uchiyama, A., Tazawa, O., Fujimoto, O., Nishiyama, E., Murase, K., Kamimura, A., Noguchi, N., Ohmura, T., Kasahara, J., 2014. Development of the Seismic ACROSS using electro-magnetic vibrator. In: 2nd International KACST-KAUST-JCCP Workshop on Surface and Subsurface 4D Monitoring, King Abdullah University of Science and Technology (KAUST), Saudi Arabia, 4–6 March.

Kumazawa, M., Takei, Y., 1994. Active method of monitoring underground structures by means of ACROSS. 1. Purpose and principle. Abstract Seismological Society of Japan, pp. 2158.

Kumazawa, M., Kunitomo, T., Yokoyama, Y., Nakajima, T., Tsuruga, K., 2000. ACROSS: Theoretical and Technical Developments and Prospect to Future Applications, 2000, Technical Report, Japan Nuclear Cycle Develop. Inst., vol. 9, pp. 115–212.

Kunitomo, T., Kumazawa, M., 2004. Active monitoring of the Earth's structure by the seismic ACROSS - transmitting and receiving technologies of the seismic ACROSS. In: Proceedings of the 1st International Workshop "Active Monitoring in the Solid Earth Geophysics", in Mizunami, Japan, S4-04, pp. 181–184.

Landrø, M., 1999. Repeatability issue of 3-D VSP data. Geophysics 64, 1673–1679.

Larsen, S., Schultz, C., 1995. A Technical Report No. UCRL-MA-121792, 18 pp.

Liebscher, A., Martens, S., Möller, F., Lüth, S., Schmidt-Hattenberger, C., Kempka, T., Szizybalski, A., Kühn, M., 2012. Überwachung and Modellierung der geologischen CO_2-Speicherung – erfahrungen vom Pilotstandort Ketzin, Brandenburg (Deutschland). Geotechnik 35 (3), 177–186.

Li, G., 2003. 4D seismic monitoring of CO_2 flood in thin fractured carbonate reservoir. The Leading Edge 22, 690–695.

Lumley, D., 2010. 4D seismic monitoring of CO_2 sequestration. The Leading Edge 29, 150–155.

Martin, Table, 1995. Cabon Dioxide Flooding. SPE, 2564.

Mathieson, A., Midgley, J., Dodds, K., Wright, I., Ringrose, P., Saoul, N., February 2010. CO_2 sequestration monitoring verification technologies applied at Krechba, Algeria. The Leading Edge 29 (2), 216–221.

Mavko, G.M., 1980. Velocity and attenuation in partially molten rocks. Journal of Geophysical Research 85, 1444–1448.

Maxwell, S., Du, C.J., Shemeta, J., Zimmer, U., Borunmand, N., Griffin, L.G., 2009. Monitoring SAGD Steam Injection Using Microseismicity and Tiltmeters. SPE, 110634.

Meersman, K., 2013. S-waves and the near surface: a time-lapse study of S-wave velocity and attenuation in the weathering layer of an Alberta heavy oil field. The Leading Edge 32 (1), 40–47.

Meunier, J., 2011. PRM techniques can significantly increase time lapse sensitivity. In: Extended Abstract of EAGE Workshop on PRM (Permanent Reservoir Monitoring) Using Seismic Data, pp. 70–73 (Trondheim, Norway).

Melzer, L.S., Davis, L., 2010. A pragmatic look a carbon capture and storage: from global issued to be technical details. First Break 28 (1), 85–90.
Misaghi, A., Landrø, M., Petersen, S.A., 2007. Overburden complexity and repeatability of seismic data: impacts of positioning errors at the Oseberg field, North Sea. Geophysical Prospecting 55 (3), 365–379.
Misu, H., Ikuta, R., Yamaoka, K., 2004. Active monitoring of upper crust using ACROSS-seismic array system. In: Abstract of Annual Meeting of Japan Geoscience Union.
Nakayama, T., Takahashi, A., Kato, A., 2008. Monitoring an oil-sands reservoir in northwest Alberta using timelapse 3D seismic and 3D P-SV converted-wave data. The Leading Edge 27, 1158–1175.
Nakstad, H., Langhammer, J., 2011. Successful North Sea fiber optic PRM installation. In: Extended Abstract of EAGE Workshop on PRM (Permanent Reservoir Monitoring) Using Seismic Data, pp. 92–93 (Tronheim, Norway).
Nagao, H., Nakajima, T., Kumazawa, Kuitomo, T., 2010. Stacking stragegy for aqisition of an ACROSS transfer function. In: Kasahara, J., Korneev, V., Zhdanov, M. (Eds.), "Active Geophysical Monitoring", Handbook of Geophysical Exploration, Seismic Exploration, vol. 40. Elsevier Pub., Netherlands, pp. 213–227.
Nur, A., Tosaya, C., Vo-Thanh, D., 1984. Seismic monitoring of thermal enhanced oil recovery processes. In: 54th Annual International Meeting. SEG, pp. 337–340. Expanded Abstracts.
O'Connell, R.J., Budiansky, B., 1974. Seismic velocities in dry and saturated crack solids. Journal of Geophysical Research 79, 5412–5426.
Paige, C.C., Saunders, M.A., 1982. LSQR: an algorithm for sparse linear equation and sparse least squares. ACM Transactions on Mathematical Software 8, 43–71.
Pevzner, R., Shulakova, V., Kepic, A., Urosevic, M., 2011. Repeatability analysis of land time-lapse seismic data: CO2CRC Otway pilot project case study. Geophysical Prospecting 59, 66–77.
Pevzner, P., Urosevic, M., Gurevich, B., Shulakova, V., Bona, A., Caspari, E., Alonaizi, F., Kepic, A., 2012. Seismic monitoring of CO_2 geo sequestration: CO2CRC Otway project experience. In: 1st Joint International Workshop for the Earth's Surface and Subsurface 4D Monitoring KACST Conference Hall, Riyadh, Saudi Arabia, January 8–11.
Richard, T., 2011. Lessons from a 4D seismic monitoring of CO_2 injection at the Delhi field. First Break 29 (1), 89–94.
Ricketts, T.A., Barkved, O., 2011. Clair permanent reservoir monitoring - a pilot that shows potential. In: Extended Abstract of EAGE Workshop on Permanent Reservoir Monitoring (PRM) Using Seismic Data, 10610, Trondheim, Norway.
Ringrose, P., Atbi, M., Mason, D., Espinassous, M., Myhrer, Ø., Iding, M., Mathieson, A., Wright, I., 2009. Plume development around well KB-502 at the in Salah CO_2 storage site. First Break 27 (1), 81–85.
Ringrose, P.S., Mathiesonb, A.S., Wrightb, I.W., Selamac, F., Hansena, O., Bissellb, R., Saoulaa, N., Midgleyb, J., 2013. The in Salah CO_2 storage project: lessons learned and knowledge transfer. Energy Procedia 37, 6226–6236.

Reuss, A., 1929. Berechnung der fließgrenze von mischkristallen auf grund der plastizitätsbedingung für einkristalle. Zeitschrift für Angewandte Mathematik und Mechanik 9, 49–58.

Robinson, N.D., Riviere, M.C., Toufsh, K., Watson, P.A., Seaborn, R., 2011. Semi-permanent monitoring of the Azeri-Chirag-Gunashili field, south Caspian Sea, Azerbaijan. In: Extended Abstract of EAGE Workshop on PRM (Permanent Reservoir Monitoring) Workshop Using Seismic Data, pp. 15–19 (Tronheim, Norway).

Saiga, A., Yamaoka, K., Kunitomo, T., Watanabe, T., 2006. Continuous observation of seismic wave velocity and apparent velocity using a precise seismic array and ACROSS seismic source. Earth Planets Space 58, 993–1005.

Sato, H., Hirata, H., Ito, T., Tsumura, N., Ikawa, T., 1998. Seismic reflection profiling across the seismogenic fault of the 1995 Kobe earthquake, southwestern Japan. Tectonophysics 286, 19–30.

Seaborne, T.R., Howe, D.J., Slopey, W., Talibov, A., Asgerov, H., 2011. Acquisition aspects of an interim life of field seismic project at the Azeri-Chirag-Gunashili field in the Caspian Sea. In: Extended Abstract of EAGE Workshop on PRM (Permanent Reservoir Monitoring) Using Seismic Data, pp. 110–114 (Tronheim, Norway).

Saito, H., Nobuoka, D., Azuma, H., Xue, Z., 2008. Time lapse cross well seismic tomography monitoring CO2 geological sequestration at Nagaoka pilot project site. Journal of MMIJ 124, 78–86.

Sakai, A., 2008. Development of the method to evaluate the distribution of the injected carbon dioxide and physical parameters by means of time-lapse of 3D seismic survey. The Japanese Association for Petroleum Technology 73, 186–199.

Schenewerk, P., April 2, 2012. EOR can extend the promise of unconventional oil and gas. Oil & Gas Journal 48–52.

Sengupta, M., Katahara, K., Smith, N., Kitrridge, M., Blangy, J.P., 2015. Modeling anisotropic elasticity in an unconventional reservoir. The Leading Edge 34 (11), 1332–1338.

Simmons, G., 1964. Velocity of shear waves in rocks to 10 kilobars, 1. Journal of Geophysical Research 69, 1123–1130.

Schissele, E., Forgues, E., Echappe, J., Meunier, J., de Pellegars, O., Hubbans, C., 2009. Seismic repeatability. In: Is There a Limit: The 71th EAGE Annual Meeting.

Sumino, Y., Anderson, O.L., 1984. Elastic Constants of Minerals, CRC Handbook of Physical Properties of Rocks, III RS Carmichael, pp. 39–138.

Spetzler, J., Xue, Z., Saito, H., Nishizawa, O., 2008. Case story: time-lapse seismic crosswell monitoring of CO_2 injected in an onshore sandstone aquifer. Geophysical Journal International 172, 214–225.

Takei, Y., 1998. Constitutive mechanical relations of solid-liquid composites in term of grain-boundary contiguity. Journal of Geophysical Research 103, 18183–18203.

Takei, Y., 2002. Effects of pore geometry on Vp/Vs: from equilibrium geometry to crack. Journal of Geophysical Research 107 (B2). http://dx.doi.org/10.1029/2001JB00522.

Takei, Y., 2005. A review of the mechanical properties of solid-liquid composite. Journal of Geography 114 (6), 901–920.

Thomsen, L., 2013. Can we use conventional seismics in unconventional resource plays? ASEG Extension Abstract 2013 (1), 1–2.

Thomsen, L., November 2015. Geophysics in a Time of Cheap Oil: Keynote Address. CSEG Doodletrain, Calgary.

Treitel, S., Robinson, E.A., 1981. Maximum entropy spectral decomposition of a seismogram into its minimum entropy component plus noise. Geophysics 46 (8), 1108–1115.

Vasco, D.W., Ferretti, A., Novali, F., 2008. Reservoir monitoring and characterization using satellite geodetic data: interferometric synthetic radar observation from the Krechba field. Geophysics 73 (6), WA113–WQ122.

Vasco, D.W., Rucci, A., Ferretti, A., Novali, F., Bissell, R.C., Ringrose, P.S., Mathieson, A.S., Wright, I.W., 2010. Satellite based measurements of surface deformation reveal fluid flow associated with the geological storage of carbon dioxide. Geophysical Research Letters 37, L03303. http://dx.doi.org/10.1029/2009GL041544.

Voigt, W., 1928. Lehrbuch der Krystallphysik. B. G., Teubner, Leipzig.

Vesnaver, A., Menanno, G., 2012. Accuracy analysis of micro-earthquake hypocenters in a CO_2 sequestration experiment. In: KACST-JCCP 1st Joint International Workshop for the Earth's Surface and Sub-Surface 4D Monitoring in 2012, Riyadh Saudi Arabia, January 8–11.

Wapenaar, K., 2004. Retrieving the elastodynamic Green's function of an arbitrary inhomogeneous medium by cross correlation. Physical Review Letters 93, 25430(1–4).

Wapenaar, K., Fokkema, J., 2006. Green's function representations for seismic interferometry. Geophysics 71 (4), SI33–SI46. http://dx.doi.org/10.1190/1.2213955, 8 FIGS.

Wapenaar, K., Draganov, D., Snieder, R., Campman, X., Verdel, A., 2010a. Tutorial on seismic interferometry. Part I: basic principles and applications. Geophysics 75, 75A195–75A209.

Wapenaar, K., Slob1, E., Snieder, R., Curtis, A., 2010b. Tutorial on seismic interferometry: Part 2-Underlying theory and new advances. Geophysics 75, 75A211–75A227.

Watanabe, T., 1993. Effects of water and melt on seismic velocities and their application to characterization of seismic reflectors. Geophysical Research Letters 20, 2933–2936.

Watanabe, T., Asakawa, E., 2004. Differential waveform tomography for time-lapse crosswell seismic data with application to gas hydrate production monitoring. In: SEG Int'l Exposition and 74th Annual Meeting Denver, Colorado.

White, R.E., 1980. Partial coherence matching of synthetic-seismograms with seismic traces. Geophysical Prospecting 28, 333–358.

White, D., 2009. Monitoring CO_2 storage during EOR at the Weyburn-Midale field. The Leading Edge 28 (7), 838–842.

White, D.J., Roacha, L.A.N., Robertsa, B., Daley, T.M., 2014. Initial results from seismic monitoring at the aquistore CO_2 storage site, Saskatchewan, Canada. Energy Procedia 63, 4418–4423.

Whittaker, S., Wildgust, N., November 2011. In: Lessons Learned: IEAGHG Weyburn-Midale CO_2 Monitoring & Storage Research Project. Industry CCS workshop, Dusseldof.

Xue, Z., Ohsumi, T., 2004. Seismic wave monitoring of CO_2 migration in water-saturated porous sandstone. Buturi-Tansa Exploration Geophysics 57, 25–32.

Xue, Z., Matsuoka, T., 2008. Lessons from the first Japanese pilot project o saline aquifer C2 storage. Journal of Geography 117 (4), 734–752.

Xue, Z., Watanabe, J., 2008. Time lapse well logging to monitor the injected CO_2 at the Nagaoka pilot site. Journal of the Mining and Materials Processing Institute of Japan 124, 68–77.

Yamaoka, K., Kunitomo, T., Miyakawa, K., Kobayashi, K., Kumazawa, M., 2001. A trial for monitoring temporal variation of seismic velocity using an ACROS system. Island Arc 10, 336–3487.

Yin, H., 1992. Acoustic Velocity and Attenuation of Rocks, Isotropy, Intrinsic Anisotropy, a Stress Induced Anisotropy (PhD. thesis). Stanford University.

Yoshida, Y., Ueno, H., Takahama, S., Ishikawa, Y., Yoshikawa, S., Katusmata, A., Kunitomo, T., Kumazawa, M., 2010. Characteristics of ACROSS signal transmitted from the Tono transmitting station and observed by Hi-Net. In: Kasahara, et al. (Eds.), "Active Geophysical Monitoring", Handbook of Geophysical Exploration, Seismic Exploration, vol. 40. Elsevier Pub., pp. 463–472

Yoshida, Y., 2011. Seismology and Volcanology Research Department, MRI, Chapter 2.3 of Improvement in Prediction Accuracy for the Tokai Earthquake and Research of the Preparation Process of the Tonankai and the Nankai Earthquakes, vol. 63. Technical Repot of the Meteorological Research Institute, pp. 115–148.

Zimmer, M., 2003. Controls on the Seismic Velocities of Unconsolidated Sands: Measurement of Pressure, Porosity and Compaction Effects (Ph.D. thesis). Stanford University.

INDEX

'*Note*: Page numbers followed by "f" indicate figures and "t" indicate tables.'

S

U

V

W

X

Z

Printed in the United States
By Bookmasters